TRAVELERS

TRAVELERS

The American Tourist from Stagecoach to Space Shuttle

HORACE SUTTON

William Morrow and Company, Inc.
New York   *1980*

Library of Congress Cataloging in Publication Data

Sutton, Horace.
  Travelers.

  Includes index.
  1. Travelers—United States—History.   I. Title.
G222.S97   917.3′04   80-14931
ISBN 0-688-03694-5

Printed in the United States of America

First Edition

1 2 3 4 5 6 7 8 9 10

Book Design by Bernard Schleifer

*For the men and women
who work in travel, with
whom I have labored
over the span of decades,
who taught me that
there is indeed a family
of man.*

Foreword

THE WAY THAT HISTORIANS count the age of a nation or a society, two hundred years is an eye's blink. In England it is the difference between Henry VIII and George III, in Russia the time between Ivan the Terrible and Catherine the Great. Viewed that way, two hundred years seems to have scant significance to anyone except a scholar. Yet two hundred years ago colonial Americans were bouncing over rutted trails in swaying stagecoaches. And when they went to Newport, the finest resort of the day, they journeyed by boat because it was the only way to go. Roads didn't as yet stretch into such rural purlieus as Rhode Island.

Two hundred years later here we are flapping five miles high across a 2600-mile-wide nation, and taking only four hours and a half to do it. Easterners spend the weekend in Acapulco, fly to Frankfurt for a meeting. Circumnavigating the globe is something any frugal workingman can do as soon as he saves up $1500, a typical fare. A million Americans go to Europe every year. Ten million take vacations in their cars. We are flying faster than sound, zipping to Paris from New York in three hours and a half. And now we are talking about space shuttles, hotels on the moon and flying wings that will carry 1700 passengers.

This is the story of how a handful of settlers trudging off to a miserable spa in the mid-1700s became, in the tick of history's clock, a mammoth industry equipped with armadas of jet planes, fleets of enormous air-conditioned ships with as-

7

sorted swimming pools, resort hotels that make snow when it doesn't fall; city hotels with rotating restaurants on the roof, glass elevators that slide up the outside wall, banquet rooms that serve dinner simultaneously to three thousand customers in the same hall, keyless guest rooms that open to a push-button code, computers that accept orders for breakfast, keep the hotel inventory and compute the guest's bill.

This is the story of the American traveler as he began to ride railroads, took steamboat rides down the Mississippi, went swimming in the sea at Old Point Comfort, the ladies disporting within the proper protection of stockades built in the water. It is the story of America in the gingerbread glory of the Victorian years, of Flagler's invasion of Florida, of the first horseless carriages and the tourist penetration of Southern California and a prediction of a reporter "that it will become the great natural health resort of this continent."

It is also the story of the war that took the Americans to Paris in 1917 and why it has been difficult ever since to keep them down on the farm or even at home. It is the story of fledgling airlines and the thirsty 1920s and the wild rush to sail off to Europe or get in a car and drive. It is the beginning of Miami Beach and the beginning of the Depression and airlines that flew coast to coast, then across the Pacific and finally across the Atlantic. It is the story of the war again in the 1940s, when Americans discovered the world beyond Paris and became infected with a restlessness and curiosity—that irrepressible combination that will never abate. It is the story of how we discovered the world, of the great hotels, the elegant resorts, of the jet people and the beautiful people and the Place-setters; of packaged tours, of how great shipping fleets turned into cruise boats, and how the propless plane that can fly to China from New York in one giant leap has made a Magellan of every American man, woman and child free to explore the universe.

HORACE SUTTON

Chappaqua, New York
1980

Contents

I | *The Colonials*

STRANGELY ENOUGH, spas, which have been violently in fashion
and hopelessly out of style in America during the last two hundred years, and are at the moment back in vogue, provided the
first form of vacation life for the early colonials. Not only did
they learn from the Indians, who had long been licking their
wounds at the natural hot water springs, but they had brought
with them the English predilection for places like Bath, Barnet and North Hall, medicinal springs which had long been
popular rookeries back in the old country.

Besides the tonic effects of the waters, the early settlers,
just like the Romans, appreciated social benefits of the baths
where one could meet and mingle while being laved. In the
early years in the colonies there was little enough to do for
amusement, and when the watering places began to sprout
along the eastern seaboard the settlers were delighted to find
a place to spend their few leisure hours. In the strict Puritan
atmosphere there was no time for aimless pleasure, but at the
spas the enjoyment of socializing was only a side product of
what was professed to be an excursion in pursuit of health.
A mission with such a sensible purpose made it all seem very
worthwhile.

As far back as 1669 Bostonians were trooping off to Lynn
Spring. About a dozen years later, when William Penn arrived
to accept his land grant, he was astonished to find mineral
waters "which operate like Barnet and North Hall . . . not
two miles from Philadelphia." Springs appeared in Virginia, in

11

Pennsylvania, in Connecticut, Massachusetts, New York and New Jersey. By boat and by coach New Yorkers were bounding off to a resort called Perth Amboy on the Jersey coast, a circumstance of history which may startle easterners who know the place as an industrial center that puffs mightily, and not always pleasantly, with chemical and paint plants, steel mills, clothing factories, printing plants and cigar companies. Bostonians patronized Lynn Spring for nearly a hundred years until the fashion of times created a new favorite, Stafford Springs in Connecticut.

If a spa could remain in favor for a hundred years it is safe enough to say that public taste was not as mercurial as it is today. Resorts these days do their best to attract celebrities. Public relations people hope to imply a testimonial in the public mind by the mere presence of stars and notables. It was the custom then to obtain a testimonial from a cured visitor or, better yet, from a physician of note. An energetic Boston newspaper traced the switch to Stafford Springs to a certain Mr. Fields who, having suffered what was described as "an obstinate cutaneous complaint," took the cure at Stafford and gave out the success story to the press. Stafford achieved an immediate reputation for curing the gout, sterility, lung problems and hysterics. When, in 1773, a group of leading doctors published a glowing testimonial under the title "Experiments and Observations on the Mineral Waters of Philadelphia, Abingdon and Bristol," it was as if Prince Charles, Liza Minelli, Halston and the Aga Kahn had all publicly announced that the Poconos were the most glorious vacationland in the world. The rush of customers approached a stampede.

Physicians who dealt in the therapy of medicinal waters were called "balneologists" and "itinerant medicasters" and, sometimes, other terms that were less polite. When the management of Stafford Springs imported an English physician who cooperated by issuing an impressive report on the quality of the waters there, some of the press, not exactly friendly to the Crown and its partisans, was inspired to launch a flight of arrows. Said a columnist in the Boston *News Letter*, "Are you for repairing to these fountains of health in Stafford . . . behold one of the royal race of Tudors will deal out salvation

to you in copious deluge, and secure you from future decay.
. . . 'Tis apparent from the great use made of the novel Glys-
termongers, we were prodigiously deficient before their happy
introduction among us."

Despite the waspish commentary from the columnists, the
spas flourished and places like Yellow Springs outside Philadelphia reported "vast concourses of people" arriving daily not
only from Penn's capital but from all parts of the country and
from the West Indies "and other foren parts." Names like
Drinker, Allen, Penn, Shippen and Mifflin, since hallowed in
the Main Line halls, were signed on the registers. Not only did
they and those of lesser celebrity enjoy the baths, but they
tried the new sport of sea bathing as well, then settled into the
Yellow Springs restaurants to enjoy fish, crabs and lobsters.

By this time Yellow Springs and Bristol Springs had become so popular and so fashionable that some Quakers suc-
cumbed to the allure and visited the springs on the Sabbath.
Protestant ministers decried the people's "immoderate and
growing fondness for pleasure, luxury, gaming, dissipation, and
their concomitant vices." When some entrepreneurs tried to
organize a lottery to build yet another spa, the clergy stomped
off in a body to urge the governor to prohibit the scheme "for
erecting public gardens with Bath and Bagnios among us."
The clergymen had the last and the compelling word. If hot
and cold baths were so necessary to good health, they said,
then proper facilities could be added to the hospitals.

Despite their enormous popularity the facilities at the spas
had nothing in common with the splendor of Roman baths.
What's more, they were difficult to get to. A stagecoach service to Stafford Springs began to operate in the spring of 1767,
and passengers were conveyed from Boston for five dollars
each. They were allowed twenty pounds of baggage, but the
baggage wagon departed sixteen days before the passengers.
John Adams, who visited Stafford to take the cure in 1771,
described the inn as "a small house within a few yards of the
spring." There, said he, "some of the lame and infirm keep."
The famed water gushed forth at the foot of a steep hill, dye-
ing everything on which it splashed a reddish yellow. "The
water communicates that color, which resembles that of rust

iron, to whatever object it washes," Adams wrote. "I drank plentifully of the waters; it has a taste of fair water with an infusion of some preparation of steel in it which I have taken heretofore. . . . They have built a shed over a little reservoir made of wood, about three feet deep and into that they have conveyed the water from the spring; and there the people bathe, wash and plunge, for which Childe [the owner] has eight pence a time. I plunged twice but the second time was superfluous and did me more hurt than good; it is very cold indeed."

There is no indication that the Pennsylvania springs were more palatable, nor more comfortable. A visitor to Warm Springs in Standing Stone Valley near the Juniata River in 1775 found one spring for bathing and another for drinking, and the whole place "an asylum for all impatient women in cases of barrenness." Among the summer guests at the time were twenty-four "professedly indisposed" including "seven unmarried Virgins of various ages." While the place seemed like an infirmary or a hospital to the observer, one Reverend Philip Filhian, the patients or guests were not all in poor health. "They must, in strong belief, at least, be indisposed," he wrote, "or they would not submit to the Inconveniences for any length of time which the situation of the place makes necessary. It is quite in the woods."

Still, some colonists found the hard trips down the primitive roads little enough hardship to endure for the reward of the pleasures of the springs. Games, masks and revels were part of the gay life at Berkeley Springs in Virginia, where Washington came to cure his rheumatic fever in 1761, and serenades were heard all night "before the houses where the ladies stayed." One visitor to the spa was so charged by the medicinal waters, a chronicler later reported, that he was inspired to break into the room of "one buxom Kate."

Washington, who took Patsy Custis, his stepdaughter, to Berkeley Springs during the summer of 1769, found little improvement in the eight years since his first visit. It took the family a week to get there from Mount Vernon and when they arrived they found their quarters in miserable condition. It was still more than a hundred years before room service, and

Washington had to comb the countryside for a baker and blacksmith. After a month they found no relief in the medicinal waters for Patsy's epilepsy, and they left. They were scarcely past the front gates when Washington's chariot broke down and he had to set out on foot to find someone to repair it. Three hours after leaving, the family had scarcely covered a mile.

A persistent if luckless traveler, Washington crisscrossed the colonies leaving a record of many short journeys and a number of mishaps. After two hours on the road from Port Tobacco, Maryland, to Williamsburg, on one occasion, Washington got what passed in those days for a flat. Unfortunately there was no Automobile Club nor mechanic to call. He wrote, "About 11 o'clock broke my chair and had to walk to Port Tobacco where I was detained the whole day getting my chair mended, no smithy being within six miles." If he had good fortune along the roads Washington could cover about thirty-five miles a day. Like travelers of the times, he used ferries often and stayed overnight at taverns, many of which were operated by retired military people. In 1751 Washington accompanied his brother Lawrence on a trip to the West Indies. Barbados had a reputation for helping people suffering from lung trouble, and Lawrence was ill.

Not even cruising in the West Indies was much of a pleasure in Washington's time. In his log he wrote, "This morning arose with agreeable assurances of a certain and steady trade wind which after nearly five weeks of buffeting and being tossed by a fickle and merciless ocean was glad'ening knews." With the sea more gentle there was time to enjoy it. He wrote, "Saw many fish swimming about us of which a Dolphin we catch'd at Noon but cou'd not intice with a baited hook two Baricootas which played under our stern for hours." He was as excited as any first-time cruise passenger by the scenes around him. "A prodigy appear'd in ye West towards ye suns setting about six p.m.," he wrote, "remarkable for its extraordinary redness."

When the ship finally landed in Barbados, Washington was struck by "the beautiful prospects which on every side presented to our view The Fields of Cain, Corn, Fruit Trees & C in a delightful Green. . . ." He found lodgings for which he paid £15 a month "exclusive of liquors and washing," and, like most travelers, grumbled about it. Vaccinations and compulsory certificates for travelers were still a long way off. Washington found himself stricken with smallpox. He recovered in a month and was shortly up and about, accepting invitations from the local gentry "including one to partake of Beef Stake and Tripe." As winter travelers continue to be today, he was pleased and exhilarated by the pleasures of the tropics. "After dinner," he wrote, "was the greatest collection of Fruits I have yet seen on the Table. There was Granadella the Sapadilla Pomgranate Sweet Orange Water Lemmon forbidden Fruit apples, Guavas E ca E ca E ca." He had to remember that a friend had asked him to buy a barrel of limes. Saddling a traveler with commissions to fulfill is no new idea.

Although travel by colonists on pleasure excursions to the West Indies was rare, the people of the island colonies in the Caribbean came often to the mainland. The Visit U.S.A. program, which did not commence in earnest until 1961, had early beginnings. The favorite of the traveling West Indians was Newport, the same Rhode Island resort still in vogue today. Although not a spa at all, it emerged as the most elegant social center of the colonies, and was never to slip from favor through two centuries. At the beginning it held its title through a stretch of eight social summers that ended only with the outbreak of hostilities with England. Invalid Antiguans were coming to Newport as early as 1729. By 1765 the governor of His Majesty's colony at Granada was referring to "the refined and polished society of Newport." The expatriate Europeans thought the summer climate was rather like Italy. Somebody called it the "American Venice," though the only resemblance was its popularity with travelers and its situation on the sea. Someone else called it the American Montpellier after the French city near the Mediterranean famed for its handsome houses of the seventeenth and eighteenth centuries. Carl Bridenbaugh, the brilliant historian who has made a specialty of the colonial period, called it "The Bath of America."

For much of Newport's early celebrity it can thank the enterprising sea captains who were then maintaining a brisk trading route between the city and ports in the South and the West Indies. They were the salesmen, the travel agents and the skipper of the ship all rolled into one. In Charles Towne or Jamaica, they declaimed with eloquence of a travelogue commentator on the pleasures of the Rhode Island port. They said very little about the voyage for getting there was none of the fun. It was a hard sea trip, the lighthouses were few, and the coast largely uncharted. The sea trip from Jamaica to Newport took about a month. Ships from Charles Towne to Rhode Island took ten days, and the sea trip from Philadelphia took four to six days. A regular packet sailed out of New York for Newport with some frequency. But the sea route was the only way to get there. As late as 1767 there was still no road from the Eastern cities leading north into New England.

In summer rich planters were Newport's most eager customers. They flooded into Charles Towne to escape the yellow fever and country fever that bred in the swamps of the Carolina tidewater, and from there they caught the sloops for the healthful pleasures of the North. Carolina sent more visitors to Newport than did any other colony. Philadelphia was second, and Jamaica third.

The Newport *Mercury* began listing the summer arrivals. The names of prominent people were set in caps, lesser lights in regular type. Among the Philadelphians were John Wharton, Clement Biddle, Gilbert Rodman, Thomas Mifflin and Josiah Hewe. Most of them were members of the Dancing Assembly in Philadelphia. The membership assured transfer of their social position to New England. Socialites hired a whole cabin on the packet and upon their arrival rented a whole house. The *Mercury* also ran advertisements which read, "To be let. A Genteel House and Furniture with a garden and pleasantly located, by Caleb Godfrey." If you were merely well off then you rented rooms, such as "Three Genteel rooms, pleasantly situated in Marlborough Street with the privilege of a garret, yard and cellar." And if you were a bachelor of modest means, then you put up at the taverns. Mary Cowley's on Church Street was for gents and it advertised "several decent rooms and beds unoccupied."

The tourist trade that began to flow to Newport was appre-

ciated by the local citizens, who were celebrated for their
hospitality and gentility. Ladies were urged to be careful of
the effects of the salt, the air and the sun on their delicate
complexions. Queens Pearl Wash Ball was guaranteed to re-
move freckle and sunburn and render skin "delicate white
smooth and soft." Surgeon dentists came to Newport, too, to
find a market for their store teeth. John Escoffier who "makes
hair cushions for ladies" was imported from Paris. France also
sent a milliner and mantua maker who "dresses ladies heads
for a half a dollar at the shop, if waited on at the home a
dollar."

The lavish spenders were the Southerners and visitors from
the West Indies. The women bought hair cushions and other
fashions, and took lessons in French, voice culture and musical
instruments. The men took fencing lessons and patronized
teachers who advertised, "will instruct young gentlemen and
ladies to play upon the flute, violin, harpsichord, guitar and
other instruments now in use."

Newport vacationers rode the chaise to Abigail Stoneman's
tea house at Middleton, where "large entertainments will be
prepared on short notice"; they watched the horse racing on
the beach; they dined on West Indies turtles prepared by the
most celebrated of all cooks, a slave called Cuffee Cockroach.
Dinner was served at two, tea at five. They danced, had a
hot toddy at eleven and headed home. They went to readings
of opera librettos. By the season of 1772 a new summer theater
was advertising "A Grand Concert of Vocal and Instrumental
Musick by the First Performers from Boston." They went on
excursions by packet to Providence, stopping at the Oyster
House on the quay for a dozen salts. It was a splendid summer
place, full of sophisticated continental ways and big-city
pleasures that delighted the Philadelphia merchants, pleased
the British gentry from the Caribbean colonies, and flabber-
gasted the Southern planters. So many southerners came that
Newport was dubbed "the Carolina Hospital."

Oddly enough, the southern capitals looked rather unsym-
pathetically upon this summer migration to the North. They
rated tourism third in lost currency, after the customs duty on
silver and payments for northern goods. But the spas and the

resorts had an enormous effect on the colonies, for it was the search for the common pleasures of a vacation that first introduced the southern planter to the Philadelphia merchant and both of them to the New Englander. Along the breeze-swept dunes of Newport and in the steaming tubs of the spas they understood their common needs and worked out their common aims. As John Adams, eager to exchange views with countrymen from other regions, was later to recall about his visit to Stafford Springs, "The journey was of use to me, whether the waters were or not."

Not all the traveling pleasures of the colonials were limited to the spas and the seaside resorts. They visited the cities too, and coming to New York was an adventure even in the early 1700s, long before celebrities publicly declared, "I Love New York."

Cato's Road House between Fifty-first and Fifty-second Streets opened in 1712 and existed until 1760. Cato, who had been a slave, picked up pin money as a cook, and ultimately saved enough to buy his freedom from his owner in South Carolina. When he came to New York he won a reputation for his okra soup, terrapin, fried chicken, curried oysters, roast duck and South Carolina milk punch. At Christmas time he mixed his eggnogs by the barrelful. Cato was such a good cook that his patrons wondered how his owner could have sold him, even to himself.

Up in the Fifties, Cato's was out in the country. Nearer the center of the city was the Sign of the Black Horse on Smith Street between Pine and Cedar. Smith Street has since become William Street in the heart of the present financial district. The tavern was famous for its drinks, especially those made with West Indies rum. An early ball held at Black Horse in January of 1736 celebrated the birthday of the Prince of Wales. It opened with French dances, proceeded to country dances. Two were invented for the occasion and named for the prince and for the princesses of Saxe-Gotha. The City Hotel, built later in the century, occupied an entire block on Broadway between Thomas and Cedar Streets. Run by two bachelors named Jennings and Willard, the City Hotel was celebrated as the best in the country. It was gay and interesting, and dances, lectures and concerts were held there. It even had a ladies' dining room.

The city life had a dozen attractions all going at the same time. The promoters toured New York, Boston and Philadelphia with a live African leopard. Down in Charles Towne a sideshow charged admission to see the phenomenon of the age, a "White Negro Girl" with gray eyes and white woolly hair. Waxworks, in style again today, were in those times a staple spectacle in any metropolitan settlement that considered itself an up-to-date city. An early lantern slide, forerunner of the movies, played in the Northern cities and was billed, rather overgrandly, as the Philosophical Optical Machine.

The vaudeville houses imported music-hall acts from London. Visitors who journeyed to the cities applauded the slack-wire artists, and a renowned lady of certain talent who suspended herself across two chairs and "with an anvil of 300 pounds on her breast suffered two men to hit it with sledge-hammers."

While the London offerings were a great success in New York, they ran into ecclesiastical headwinds in Philadelphia. Philadelphians, when they sought a high time on the town, traveled to New York. Among other things New York was a great race town. In the world of 1750 its Newmarket Track at Hempstead Plain was the place to be. The day before the races seventy chairs and chaises took the ferry across the East River and headed for Long Island. By race time there were an estimated one thousand horsemen on the plain. Small as they were, the cities of colonial America were the exciting places to be, and New York was the biggest adventure of all. The problem was getting there, for there was only the packet, the horse or the stagecoach.

The timetables of the eighteenth-century stagecoaches paid no more heed to passenger comfort than do the transportation people of today. Indeed, perhaps the stagecoach dispatchers left a legacy. The stage for Manhattan along about mid-century was leaving Philadelphia at 4 A.M. A party of some twenty friends who made the trip recalled traveling along the King's Highway marveling at the people "all strangers in their manners to us young travelers." They might have made use of one of the first guidebooks, *Vade Mecum for America, or a Companion for Traders and Travelers*. It listed roads and taverns from Maine to Virginia and contained a directory of Boston streets in 1732.

One of the best records of travel in the colonies was left by Dr. Alexander Hamilton, of no apparent relation to the founder of the republic, whose papers were edited by Carl Bridenbaugh. Dr. Hamilton began an odyssey in Maryland in May of 1732 traveling clear to York, Maine, in search of health and amusement. Whether his health improved was never recorded, but he certainly was never without diversions. Accompanied by his slave Dromo, he spent a week in Philadelphia, continued on to New York for six rousing days in Manhattan, took a side trip by boat to Albany, and returned two weeks later for a tour of New England by way of Long Island. He was back in autumn exploring Connecticut's shore and was ready to write his comments. They were candid if not juicy. Hamilton found the northern parts of the colonies much better settled than the southern and the people "of a more gygantick size and make." In matters of politeness and humanity both sections of the country were similar, although he couldn't resist pointing out that the inhabitants of the great towns are more civilized, "especially at Boston." He was less charitable about Philadelphia. To a visitor from abroad the City of Brotherly Love of 1744 looked like this:

> At my entering the city I observed the regularity of the streets but at the same time the majority of the houses mean and low and much decayed, the streets in general not paved, very dirty and obstructed with rubbish and lumber but then frequent building excuses that.

Philadelphia in the 1740s was undergoing a general facelifting, in many ways, a situation not unlike New York in the 1960s. June in Philadelphia was warm and Hamilton decided that the excessive heat derived from the sun's rays "being reflected with such power from the brick houses and from the street pavement which is brick. The people commonly use awnings of painted cloth or duck over their shop doors and windows and at sunset throw buckets full of water upon the pavement which gives a sensible cool."

In a city famed for its merchants, Philadelphia offered the best and largest market in America, held Wednesdays and Saturdays on, naturally enough, Market Street. Hamilton found only one public clock in the city but even that was bereft, he said, of "index or dial plate" and the traveler concluded that the Quakers shun adornment. The old Philadelphia joke about rolling up the sidewalks on Sundays had early beginnings, for even in 1744 this visitor commented, "I was never in a place so populous where the goût for publick gay diversions prevailed so little."

New York was different. It made a fine appearance for a mile along the river and "here," commented Hamilton, "lyes a great deal of shipping." He concluded "By the stirr and frequency upon the streets" that it was more populous than Philadelphia. There was shipping in the harbor, the houses more compact and regular, "most of them after the Dutch modell." He found water sold on the streets, the vendors transporting it by sled. New York's women seemed more gay, appeared more often in public than the Philadelphia belles. The time to see them was in the cool of the evening when they set about on promenades, carrying their painted umbrellas adorned with feathers.

On his cruise up the Hudson, Hamilton passed Nutting Island, called that because the trees there bear nuts. Its name was later changed to Governors Island. He passed a little village called Greenwitch. It comprised eight or ten neat houses, and would later become a section of Manhattan called Greenwich Village. Albany, his river destination, was something of a disappointment. The people were all Dutch, and their way of life was quite different from the English. He was later to write, after visiting the most famous American resort, that "Newport is as remarkable for pretty women as Albany is for ugly ones." On his return trip down the Hudson he found no one to talk to. In 1744 all the passengers, on what was the progenitor of the famed Albany night boat and the Hudson River Day Line, spoke Dutch.

It was an engaging life in the colonies in the years before the war, and if the hinterlands were still a frontier there was a certain sophistication in the cities. The spas were as good for the soul as they were for the body and Newport was, of course, the last word in elegant summer living. Brought together at the watering places, the colonists better understood each other's problems and when the showdown came they were of a single mind and purpose. In June of 1775 the last social notice appeared in the Newport *Mercury*. It listed the arrival of the sloop *Friendship*, which had come up from Charles Towne with four North Carolina vacationists. When the passengers had disembarked, the vessel was taken under the protection of men-of-war. That same month Washington was named commander in chief of the Continental forces. He took command on July 3. The war was on, and the brilliant period that existed before the emergence of the Republic came to an end.

II | *The Restless New Country (1800–1860)*

THE WAR WAS OVER and the country was established. Now the new nation grew restless within its borders along the Eastern seaboard. To link the cities and transport people and goods, canals were cut and turnpikes laid out. The country grew feverish in its rush to buy shares in the new highway and bridge companies, the first real chance for stock trading in the United States. The greatest of all turnpikes and the only one financed by the government was the National, or Cumberland, Road. As early as 1802 a congressional committee recommended that a highway be build from Cumberland, Maryland, to Wheeling, West Virginia, but work wasn't begun until 1808. After it was started it took three presidents, ten Congresses, fourteen governmental acts and nine years to get the highway opened as far as Wheeling. The word boondoggle hadn't been invented yet, but the political tradition for it had gotten an early start.

Foreign visitors who came to see the new experiment in democracy were always surprised to see how few private coaches trundled over the roads. But public stagecoaches were popular and a number of competitive transport companies traversed the National Road, among them the June Bug Line, the Pioneer Line, the Good Intent Line, the Oyster Line, which carried oysters, and the Shake Gut Line, which lived up to its name, carrying packages and perishable freight. No better then than now, transportation schedules were hardly arranged to fit the convenience of the traveler, and it happened that a dozen coaches of different lines would rumble out of Wheeling at the same time in a Ben-Hur dash to be first at the relay station and get the best of the fresh horses. It took a full twenty-four hours for the fastest coach to cover the 130 miles between Cumberland and Wheeling. Anything faster would place an extreme strain not only on the horses but also on the gut-shaken passengers. Cumberland to Wheeling cost $8.25, and you could get from Baltimore to the Ohio River for $17.25. By 1802 there was through coach service from Boston to Savannah, a distance of 1200 miles. The coaches could bound over some 50 miles each day. Boston to New York required four days' journeying and cost $10. New York to Philadelphia took a day and a half and cost $5. Philadelphia to Charleston was the longest haul: it required fifteen days and the fare was $50. All in all, Boston to Savannah was a three-week run, and with board and lodging along the way, would leave little change from $100.

In 1787, four years after the Revolution had ended, a Maryland citizen named John Rumsey tried out a steamboat on the Potomac. It moved by forcing water out of the stern, a system not dissimilar from the jet-propelled yachts produced in the 1960s. Rumsey's model roared over the water at four miles an hour. He got a permit to navigate the byways of New York, Virginia and Maryland, and a syndicate with Ben Franklin as a member was formed to promote the device. Robert Fulton's 150-foot-long *Clermont*, launched in 1807, loaded its bins with pine wood and panted up the Hudson to Albany in thirty-two hours, a rather considerable achievment when you consider that his craft was about as long as the cruise boats that circle Manhattan Island today, and the distance was 150 miles. A steamboat rather hopefully christened the *New Orleans* was launched in Pittsburgh in 1811, and leaving there in October of that year made the city for which it had been named in January of 1812. Seven years later, the *Savannah* added paddle wheels to help its sails and raced from Georgia to Liverpool in 29 days and 11 hours. Technical progress was advancing at a dizzying rate. (In the air, nearly a quarter of a century was to pass between the Wright brothers' first successful flight at Kitty Hawk and Lindbergh's solo hop across the

Atlantic. Regular transatlantic passenger schedules didn't begin until just before World War II, nearly forty years after the
Wright brothers' experiment.)

Railroads began, as they had in England, with horses drawing carriages over wooden rails. A freight carrier began running from Quincy, Massachusetts, to the Neponset River south
of Boston in the 1820s, and the Delaware and Hudson Co.,
hitherto interested in moving barges along canals, laid rails
between Honesdale, Pennsylvania and Carbondale, fourteen
miles away. The Portsmouth and Roanoke took ads in the
Norfolk newspapers in August of 1834 to announce twice-a-
day service with horses pulling carriages mounted on rails.
Several weeks later the first locomotive arrived, and horses
were replaced by steam.

Even before it had a chance to lay a network of transportation, the United States, just fourteen years after Washington
first took office, was able to sign the Louisiana Purchase, which
enlarged its real estate by 140 percent. Especially important
at that time, the Louisiana Purchase brought New Orleans,
with its flamboyant ways and European culture, into the
young, severe, no-nonsense nation still properly laced in
Puritan bonds. It was like introducing a star of the music halls
into membership in the ladies' church circle. For half a century or more, visitors from the North and from abroad would
never cease to marvel at the profligate metropolis on the delta
where it wasn't a sin to have a good time. An early exponent
of Gaul *über Alles*, a French royalist named Baron de Montlezun, visited the city in 1817. He wrote:

> On disembarking in New Orleans, it is immediately ap-
> parent that one is in a French city. . . . The morose
> habits of the sad Americans have not won out here.
> Sunday is a day of pleasure as in France. The balls are
> attended with inconceivable avidity. The passion for
> dancing is at its height. The carnival is about to begin.
> It is a time for dressing up, for games, for love. The ladies
> who live in the country come into the city to take part
> in the gaiety and to contest the prize for beauty for
> elegance and for grace.

Nothing so frivolous was afoot in the original colonies. The wealthy gentry returned to the spas which they had frequented before the war, but they went, as they had before, professing indispositions which only the tonic waters or the sea air would cure. The nearest resort to New York was Bath in the town of New Utrecht, nine miles south of Brooklyn. It offered shade trees on a lawn overlooking the bay, and new accommodations a short way from the beach. A syndicate of seventy or eighty gentlemen from New York built the Marine Pavilion Hotel at Rockaway, a new resort with a fringe of fine beach. The hotel was 230 feet long, three stories high, and a "piazza" twenty feet wide ran along its entire length. It was all comfortable enough, but to get there from New York, twenty miles away, one had to take a stagecoach and a train.

Explorers wishing to divine the mysteries of New Jersey could journey to Long Branch, thirty miles from New York, or even to Cape May, which came into style after the War of 1812. By 1816 Cape May had built Congress Hall Hotel. Its grand arched dining room, two thirds the length of a football field, would be receiving guests for the next seventy years. Long Branch, sitting alongside the sea just thirty miles from New York, began soon after the Revolution's end to attract the gentry down from the cities to improve their health. Coming from New York, travelers took the boat to the Shrewsbury Inlet, then changed to stagecoach. The road to the beach was so soft that salt meadow grass had to be spread over the pathway to keep the eight-inch-wide wheels of the stage from sinking to their hubs. People from Philadelphia traveled by stagecoach all the way, and the trip was so arduous that the first few days were normally spent resting up from it.

In some resorts both in Europe and in America, where the new idea of sea bathing was becoming popular, women entered the ocean aboard bathing machines, a contraption rather like a bathhouse on wheels pulled by horses. Inside, in privacy, the ladies could change their clothes and then, once the horses had pulled them away from leering eyes, they could, in continued privacy, dip a toe into the deep. Unfortunately, the surf at Long Branch didn't lend itself to the use of these rigs and the resort was forced to devise a plan that would satisfy the requirements of Puritan propriety. As one visiting British lady put it, "Custom did not authorize *tête à tête* immersion." High on a bluff overlooking the beach, the municipality erected a flagpole. When a white flag flew the ocean was reserved to women. When the red flag went up the women retreated and the men came down from the bluff for their turn. For some reason custom also forbade women from appearing on the beach before six in the morning.

Meals began with grace, evenings were occupied with prayer meetings and hymn singing. During the days the ladies collected shells or dried starfish, which they wore on ribbons around their necks. They collected seaweed, which they dried so it could be hung as decoration over their doorways back home. It was a sedate, wholesome, pure and physically, anyway, invigorating life for the Philadelphia folk who made up the majority of the visitors. The circumspect atmosphere lasted at Long Branch for nearly ten years. Not until the 1830s did the first faint, beguiling essence of gaiety come wafting in to mix with the sea breezes. Soon men began to play cards and billiards, to bowl and race horses on the beach. Ladies and their escorts danced. Those who rued this new dissolute atmosphere clucked their tongues and moved out of the beachfront hotels to stay with the farmers.

As the country expanded and gained new maturity, new territory and new immigrants, the Puritan strain on which it had been founded, and which had given it such a dour reputation among visiting Europeans, now began to thin out. While travelers hustled off to vacations to the seaside establishments at Long Branch, Cape May, Nahant in Massachusetts, and Newport, which had been popular even before the Revolution, they also bounced off by rump-springing stage to the spas that stretched as far north as Poland Spring in Maine and as far south as White Sulphur in Virginia. In between were Pennsylvania's Yellow Springs, Connecticut's Stafford Springs and the medicinal waters that bubbled out of Schooley's Mountain at Hackettstown, New Jersey.

But the popular spa of the emerging era of libertinism was Saratoga, a fountainhead of 150 springs and wells in the foothills of the Adirondacks, thirty-three miles north of Albany.

Although the tonic qualities of its waters, curative of nearly any ill from hangnail to hives, were first noted in the late 1700s, it was the building of the Erie Canal together with the understanding ways of its hotelkeepers that were responsible for its first success. The hotels countenanced—and, it is darkly said, even encouraged—such licentious practices as drinking, dancing, laughter and unchaperoned afternoon drives through the greening woodlands. Moneyed New Yorkers flocked to it, and early in its long career, it had a reputation for being glamorous and sophisticated, if not a little racy. When the railroad followed the crowds and opened service to Saratoga in 1833, it was a hard blow to competitive wateringholes. Only Cape May, which didn't get a railroad until 1863, gave it real competition, and if anything attracted a crowd that was, socially perhaps, more fastidious.

Until the 1840s gambling had been tolerated at Saratoga as long as it was confined to friendly games held in private rooms. Later it boldly emerged into public view. Billiard rooms were built into the hotels, and gambling houses and bowling alleys appeared. Visitors flocked to Saratoga from the South and West, and it became a place of fashion.

Saratoga built big hotels, among them the celebrated United States, Congress Hall and Union Hall. The traditional hotel architecture of the day called for rows of stalwart columns not only to please the eye, but to support the roof which extended over the broad verandas. Aside from gambling and taking the waters, the regimen required an after-dinner promenade in which all of Saratoga's assembled guests, as if on command, would march in review inspecting each other as if with lifted lorgnette. It was a good way to find a new face or a new fashion. The possibility of unearthing an eligible prospect for matrimony was not overlooked either.

The real reasons for coming to Saratoga—the baths and the waters—were altogether forgotten. It was claimed, as a guide-book of 1821 put it, that Saratoga's waters were helpful in cases of "Jaundice and billious affections generally, Dyspepsia, Habitual Costiveness, Hypochondrical Complaints, Depraved appetite, Calculous and nephritic complaints, Phagedenic or ill-conditioned species or states of gout, some species of dropsy,

Scrofula, paralysis, Scorbutic affections and old Scorbutic ulcers, Amenorrhea, Dysmenorrhea and Chlorosis." Baths cost a quarter a go in the public bathhouses, and you could get five tickets for one dollar. Soap and towel were included.

In the course of giving the young nation an improved constitution, Saratoga also gave it the Saratoga trunk and the Saratoga chip. The trunk was large, edged in iron, curved on top, filled with fashions, and the bane of life to expressmen, porters and bellhops who had to heft it. The chip is said to have been invented in the momentous year of 1853 at Monn's Lake Hotel by a sarcastic chef called George Crum who, fanciful history says, was the son of a mulatto jockey and an Indian girl and had five wives all of whom worked as waitresses at the hotel. When a guest at Monn's sent back Crum's french fried potatoes and requested that they be sliced a bit thinner, Crum shaved off a paper-thin sliver, dropped it in grease and had the result carried into the dining room. The customer wasn't angry, he was delighted, and a new delicacy was declared invented.

Although Saratoga had built big hotels—the first one had gone up as early as 1802—comfortable accommodation was its biggest problem. Visitors compared Saratoga hotel furniture with the decor found in jails. William Seward, then governor of New York (and later to be Lincoln's Secretary of State), visited Saratoga in 1842 and found his suite supplied with such inadequate furniture that he was forced to descend to the public rooms to write a letter.

Southerners of lofty rank and station often went to White Sulphur Springs, where they fared no better. A British visitor who visited the Virginia springs in the late 1830s wrote:

. . . the dinners and other meals are, generally speaking, bad; not that there is not a plentiful supply, but that it is difficult to supply seven hundred people sitting down in one room. In the morning, they all turn out from their little burrows, meet in the public walks, and go down to the spring before breakfast; during the forenoon when it is too warm, they remain at home; after dinner they ride out or pay visits, and then end the day, either at the ball-room or in little societies among one another. . . . The crowd is very great, and it is astonishing what inconvenience people will submit to rather than not be accommodated somehow or another. Every cabin is like a rabbit burrow. In the one next to where I was lodged, in a room about 14 feet square, and partitioned off as well as it could be, there slept a gentleman and his wife, his sister and brother, and a female servant.

The manager of the springs was said to be a despot, nearly dizzy with power, for despite the uncomfortable quarters, the cream of Southern society flooded here. Known by visitors as the "Metternich of the Mountains," the majordomo denied entry to the unwanted and the unknown, and generally speaking admitted only those who traveled in their own private carriages. Foreign visitors were somewhat perplexed to see how aristocratic and selective and socially conscious America could be, despite the fact that it had been founded on principles of democratic equality. Said a visitor from abroad, harking back to the famed Belgian resort which had given its name to all watering places, "Spa in its palmiest days, when princes had to sleep in their carriages at the doors of the hotels, was not more in vogue than are these white sulphur springs with the elite of the United States. . . . I certainly must say that I never was at any watering place in England where the company was so good and so select as the Virginia springs in America."

No less a poet than Francis Scott Key recorded some of the ills of the springs during a visit in 1830. His verses, which were scribbled in the diary of a Virginia belle, went like this:

There's an insect or two called a flea, that here stings
The skins of the people who stay at the Springs;
There's a broom and a half here for nobody brings
Such articles here to sweep out the Springs;
There's bawling all day—but the ball at night clings
The most to my fancy here at the Springs—

To conclude, though some things here might e'en do
   for kings
If you wish to fare well, say farewell to the Springs.

Twenty years later White Sulphur Spring was even more popular and just as unbearable. Arrivals were pouring in at the rate of a hundred a day, and the most beautiful girls of the Southland were, as one visitor put it, "holding their court of love at this fountain." Generals, judges and colonels—but few of lower rank than that—flooded into the springs, but no title, however high, would save a single man the discomfort of sharing his room with another guest. Indeed, if he had a room he was lucky. The floors of the drawing rooms and parlors were strewn with mattresses.

Once they secured some sort of accommodation, guests still had to worry about being fed. One visitor noted that even when you give the waiters extra tips they bring you nothing. Cried he:

Poor fellows, they have nothing to bring! for the flour has given out; the cows have been milked dry; the mutton has run off into the mountains; and the chief cook has gone distracted! If you can manage to seize upon a bit of beef, and a slice of bread, 'tis your main chance, and hold on to it. Do not run any risks in looking about for vegetables, much less for side dishes, or pepper, or salt. For, while you are vainly endeavoring to accomplish impossibilities, some light-fingered waiter, under the pretence of changing your plate, will run off with your only chance of a dinner.

The scene presents a most ludicrous struggle for bones, and cold potatoes. Or, rather, it is fearful to witness such a desperate handling of the knife; to see so many faces red with rage at getting nothing; and ladies' cheeks pale with waiting; and starving gourmands looking stupefied into the vacuum of the platters before them. . . . Truly, the Frenchman who dines on the hair of his mustache, and the end of his toothpick, in front of the *Cafe de Paris*, is a lucky fellow, and has something under his jacket, compared with these boarders at two dollars per diem.

But it is still worse dining, when it rains. The ancient roofs of some of these halls and piazzas are not made of caoutchouc; and you cannot then sit at meat without two black boys at your back—one to keep off the flies, and the other to hold over your head an umbrella. There is a good excuse for the soup being thin on such days. 'Tis, in fact, mere rain water, with, possibly, a fly, or two, in it.

Despite the discomforts which visitors were forced to endure because of the crowds that rumbled over the rutted mountain roads to be at the watering places, few would have enjoyed it if the spas had been empty. "The more colonels the better," one visitor wrote. "The more pretty ladies the gayer. Half the pleasure is in the excitement which proceeds from the great number of persons collected together. Let the fashionable crowd dwindle down to a few dozens and you leave also. Then you can have an entire suite of rooms, and excellent dinners, with a waiter at each elbow. But no. When you see the trunks brought down, and hear the farewells said, you are as homesick as anybody and crowd into the ninth place in the coach, rather than run the risk of being the last man to leave the mountains. . . ."

The rush to the springs started with the advent of the summer season. Ever since colonial times the southerners, particularly in the coastal towns, retreated inland to the cool springs in the hills, where they could be out of reach of the heat, the malaria and what they called the "miasma." Here is how a visitor to the Virginia springs in the 1850s assessed the ills, the cures and the assorted local waters: "The dyspeptic is put on alum water, and the Southwesterner, with bile in his blood and jaundice in his eyes, is ordered to drink of the White Sulphur or the Salt. The Healing Spring is good for the gout; ladies, weary after the winter's dancing, are strengthened by bathing in the two Sweet Waters; the Blue Sulphur, taken before eating venison steaks, is said to be excellent against all devils of the same color . . . the Red Sulphur is a certain cure for consumption."

A travel writer for Raleigh's *Southern Weekly Post* wrote,

> "The Spring Season" is almost at hand, and soon the
> moneyed part of our population who have time at their
> command, will be moving off on various lines of travel,
> in pursuit of health and excitement at the numerous
> watering places and fashionable resorts of the Union.
> Newport, Saratoga, Niagara, Cape May, the Virginia
> Spring, Old Point Comfort, Nag's Head and many other
> places of less note will be thronged with visitors, and
> almost everyone of these will have some plea of ill health,
> some dyspepsia or rheumatism or other, to reveal to their
> acquaintances as the true cause of their travels.

Although one might have expected to find a collection of
the crippled, the afflicted and the infirm in residence at the
spas, the prime diversions, especially at the Southern spas,
were flirting and lovemaking. Southerners were fond of sitting
in the shade and talking for hours. Northerners rose with the
sun and climbed over the mountains in search of exercise.
But everyone looked forward to the day's most important
event, the arrival of the stagecoach.

Of all the ways of getting around in America during the
early 1800s, stagecoach was the least desirable. An English
officer traveling between Richmond and Charleston on the
mail coach early in the century, called his experience "no light
service." The coaches broke down on an average of twice a
week, and "the wrecks and the maimed" were always found
on the roads. Baron de Montlezun, the French nobleman who
had found New Orleans so enchanting, once wrote, "I must
repeat again and again that the American stage coaches are
untrustworthy, and often an insult to common sense . . . to
pass from a steamboat to a stage, especially in bad weather,
is to descend from paradise to hell." The famed French writer
and liberal politician de Toqueville, who wrote a classic piece
of exposition on American democracy, also had some candid
words to say on his travels inside the nation. Traveling between Albany and Utica in the 1830s, he called the road "in-
fernal," noted the carriage was "without springs or curtains."

The calmness of the Americans about all these annoyances, he found extraordinary. "They seem to put up with them as necessary and passing ills," he wrote. "The wealthiest people travel in public conveyances without a serving woman and without luggage. Generally speaking, they never complain about anything and adapt themselves to all the discomforts of travel in a way that shows that their education is different from that received by women of the best circles of Europe."

Charles Dickens, who visited America in 1842, once took a four-horse carriage from York, Pennsylvania, to Harrisburg. It carried twelve people inside and a large rocking chair and a dining table were strapped to the roof. Riding into Fredericksburg, he found the American coaches reminiscent of the French but not "nearly so good because instead of springs these are hung on bands of the strongest leather . . . they are covered with mud from the roof to the wheel tire, and have never been cleaned since they were first built."

The English novelist Anthony Trollope was taken to America when he was a boy by his mother, Frances Trollope. She traveled the breadth of the country and then returned home to write a controversial book called *The Domestic Manners of Americans.*

> The stages [she wrote] do not appear to have any regular stations at which to stop for breakfast, dinner and supper. The necessary interludes, therefore, being generally impromptu, were abominally bad. We were amused by the patient manners in which our American fellow travelers ate whatever was set before them, without uttering a word of complaint or making any effort to improve it, but no sooner reseated in the stage than they began their complaints—"'twas a shame"—"'twas a robbery"—"'twas poisoning folks"—and the like. I, at last, asked the reason of this and why they did not remonstrate? "Because, madam, no American gentleman or lady that keeps an inn won't bear to be found fault with."

Although Americans clucked disapprovingly at Mrs. Trollope's severe criticism of them and of their rather poor facilities for the traveler, the nation got no better editorial treatment at the hands of its own citizens. An American writer traveling the government-constructed National Road in 1833 found the ruts worn "so broad and deep by heavy travel that an army of pygmies might march into the bosom of the country under the cover they would afford." Still, inns and taverns were spaced about twelve miles apart, and it was said in a maxim, probably advanced by the proprietors of the stagecoach lines, that the coaches were so regular that the farmers could tell time by their passing. Nobody, not even their owners, said the ride or the road was smooth. In 1835 no less a personality than John Marshall, Chief Justice of the Supreme Court of the United States, suffered severe contusions on a ride from Washington to Richmond through the flat Virginia tidewater. He died the same year.

On America's waterways there was less chance of injury or discomfort, at least physically. The canal-building balloon burst with an ear-splitting explosion in 1837 and many were maimed financially. Still, during the prebust years, canal travel was a slow but smooth way of travel, the horses pulling the barges along the towpath. Some vestiges of the system remain, albeit for the pleasure of tourists, in New Hope, Pennsylvania. In southern India horses still pull freight boats along towpaths. In America, during the height of the canal days, the trip from Albany to Buffalo took a day longer than the train, but it cost only $7.20 including meals. Special rates were posted to lure the immigrants away from the railroads, which were running special immigrant cars.

Dickens wrote of taking a canal boat from Harrisburg to Pittsburgh. It was little more than a barge with a house on it pulled by three horses with a towrope. Ladies' and gentlemen's quarters were divided by a red curtain, and shelves came down at night for sleeping. Dickens described them as "just the width of an ordinary sheet of Bath post letter paper." The trip took two full days and nights.

The New York–Albany boat had two cabins, one for men and the other for women, and the gentlemen were not permitted on the ladies' side without permission from the ship's officers and the concurrence of the entire passenger list of ladies. Despite the fact that the packets carried as many as 300 passengers, dinner was handsomely served; and when the cabin was lighted later in the evening for a snack of tea and sandwiches, it rather looked like a ballroom supper. The cost of the trip was seven dollars, meals included. Stewards did not take tips.

Separation of the sexes was nowhere more rigidly maintained than on the Mississippi boats, where even husbands and wives were billeted apart, meeting only at mealtime. Aboard the *Lady Franklin*, a renowned ship putting in at Cincinnati and Wheeling, the ladies spent their days lolling under an awning spread over a large balcony in front of the ladies' cabin. "Of the male part of the passengers," Mrs. Trollope wrote, "we saw nothing, excepting at the short silent periods allotted for breakfast, dinner and supper at which we were permitted to enter their cabin and place ourselves at their table." Dickens, who journeyed from Washington to Potomac Creek en route to Richmond aboard a steamship, noted that the washing and dressing "apparatus," as he called it, "consists of two jack towels, three small wooden basins, a keg of water and a ladle to serve it out with, six square inches of looking glasses, two ditto ditto of yellow soap, and a comb and a brush for the head, and nothing for the teeth. Everybody uses the same comb and brush except myself. Everybody stares to see me using my own, and two or three gentlemen are strongly disposed to banter me on my prejudices, but don't."

The Albany boat bound for New York Dickens found "so crowded with passengers that the upper deck was like the box lobby of a theater between the pieces, and the lower one like Tottenham Court Road on a Saturday night. But he called his Lake Champlain steamer "a perfectly exquisite achievement of neatness, elegance and order. The decks are drawing rooms; the cabins are boudoirs, choicely furnished and adorned with prints, pictures and musical instruments; every nook and corner in the vessel is a perfect curiosity of graceful comfort and beautiful contrivance."

By the late 1840s steamboat travel on the Great Lakes,

too, had become just as elegant. The boats carried barber-shops, the dining salons were staffed with efficient Negro waiters and the food and service were as good as the best hotels. The cost of traveling by boat and staying in a good American hotel were about the same. Buffalo to Milwaukee, a four-day trip, cost $10.

Travel on steamboats rose enormously. In 1841 St. Louis tallied 143 steamboat arrivals from the Upper Mississippi. Five years later the number was up to 663. Northbound boats from Chicago were jammed, and beds were often made up on the floor for those who couldn't find a stateroom. There were three table settings and since the last guest got what was left from the onslaught of the first two, the gents arrived at the tables an hour before mealtime, and arrayed themselves before the empty plates to wait.

The panic of 1837 shook the young nation the same way that the older nation was to be rocked by "Black Friday" of 1873 and the mature nation would later be stunned by the market crash of 1929. The speculative fever to build canals and buy canal stock was an epidemic that proved fatal to the patient; the era of canals began a slow and agonizing death. When the smoke of '37 had cleared, the steam railroad had replaced the canals as a means of locomotion. The highway promoter and the canal companies had fought hard to suppress the railroads, and so, for that matter, had the people who thought it plain sinful to rocket along at fifteen miles an hour. It was perfectly clear, they said, that if the Maker had wished man to move faster, he would have provided him with something better than legs.

Despite these obstructions, the railroads advanced, even though the early passengers were alternately asphyxiated, choked, sprayed with soot, showered with sparks and other-wise inconvenienced. The earliest cars offered no ventilation except the ordinary side windows of the coaches. When they were opened, hot ashes from the engine, black soot from the stack and dust from the roadbed poured in on them. When they were closed the passengers gasped for breath. Having developed the steam engine, railroads had by 1836 finally invented a way to ventilate the carriages. They cut holes in the roofs of the carriages and covered them with tin when it rained.

Soon trains began to run at night. Passengers who wanted to sleep simply lay down on the hardwood benches. The pressure of competition from the canal boats and from the steamboats, which made special and, in some cases, rather elaborate provision for sleeping travelers, finally forced the railroads into action. They responded by arranging three collapsible palettes which were lowered to make horizontal bunks. Eventually the railroads were moved to provide mattresses, which were stored during the day in a pile at the end of the car. Seats fitted with springs and upholstered made their first appearance in the 1850s. Still, there were no classes of travel excepting for Negroes and immigrants, each of whom was set apart in separate cars. In New York companies arranged for the mass transport of immigrants, who piled into the immigrant cars with all their belongings. But often the companies cheated them by dropping them off halfway to their planned destinations. The cheaters must have realized the narrowest of profits, for the reliable companies offered transport clear to Wisconsin for $6 with no payment required until the immigrant had been deposited at this agreed destination.

Regular passengers paid about three or four cents a mile and, as one English traveler put it, "Although their fares appear low, their accommodation is on a par with it." However, you could check your baggage at the depot and upon arrival pay twenty-five cents to have it delivered directly to your hotel room. Railroad meals were a standard fifty cents, but the food was less uniform. As one New Hampshire traveler put it, "A man might as well get a quid of tobacco with his money, for he seldom gets a quid pro quo." After an all-night ride, the train would stop and the conductor would allow twenty minutes for breakfast. Anybody lucky enough to get a waiter, let alone something to eat, within that allotted time would find himself served with a cup of black coffee ("strong of water" was the New Hampshireman's assessment of it), a tough fried beefsteak, some fried potatoes, a heavy biscuit, a little sour. "Everything is sour," the New England man reported, "but the pickles." It was not until 1858 when an outraged passenger traveling on the Lake Shore Line between Buffalo to Chicago decided that something had to be done to alleviate the discomfort of the man on the road. His name was George Pullman and he finally produced his first sleeping car, the *Pioneer*, in 1864.

But the century was well more than half over and the Civil War was almost done by the time the sleeping car emerged. Americans and foreigners who came here on visits early in the century weren't dissuaded by the rigors of the road. Travelers bounced about from city to city, exclaiming over the marvels they found. Davy Crockett, a congressman from Tennessee during the early 1800s, toured the North and the East and published a travel book about it, although there are doubts now about its authenticity. Two years before he perished at the Alamo, Crockett looked on New York for the first time and pronounced it "a bulger of a place." It had thirty hard-pressed hotels and the overflow slipped into the boardinghouses scattered through the city. Largest and most famous of all the hotels was the Astor House, an enormous granite fortress six stories tall that contained six hundred beds and was said to cover a greater area than any hotel in London or Paris. Built by John Jacob Astor on the corner of Broadway and Vesey Street in lower Manhattan, it filled the block from Vesey to Barclay Street and looked down from its eminence upon City Hall Park. With its Doric columns standing alongside its main entrance on Broadway, its rotunda a favorite luncheon place, its air conservative, it was Broadway's fanciest pile. It was a smashing forerunner to the string of hotels which the Astor family went on to build and it lasted right up to the edge of World War I.

A week's lodgings at places like the Custom's House Hotel, Tammany Hall and Lovejoy's cost anywhere from $2.50 to $3.50 a week, and a one-plate meal in their restaurants cost from 12½ cents to 31½ cents. The idea of a hotel that charged only for lodging and gave the guest the option of eating where he liked was imported from Europe in the early 1830s. By 1837 there were three hotels using this European plan, but the others offered an all-inclusive or American plan price with a long table, or *table d'hôte*, as the French called it, set up

each day. The bed and board, or American-plan rate, cost
$1.50 a day at some places, and as much as $2.50 at the Astor
House.

Later came the St. Nicholas, which one British visitor at
mid-century point was moved to call "the greatest hotel in
Christendom." It could offer shelter to a thousand travelers,
whom it pampered with marble-paved smoking rooms, a billiard room with ten tables, a newsroom that proclaimed bill-
board announcements of the comings and goings of the Russian
Court or the prices of slaves on the New Orleans market. A
bar served such American concoctions as the brandy smash,
mint juleps and gin slings. A large and elegant hotel, the
St. Nicholas was entitled to charge $2.50 a day, which included
everything but wine. Naturally, it had a separate entrance for
women. The main entrance, crowded with Yankee smokers,
spitters and tobacco chewers, was, as a London visitor noted,
no place for a lady.

There were three theaters in New York by 1830, six by 1837,
most of them filled nightly. Some of the proper gentry looked
with disdain on the theaters because neither the Sabbath nor
Christmas Day inspired them to close their doors. Broadway
disappointed some visitors because it was not as wide as
its name implied, nor were all its buildings as imposing as
Stewart's department store, the Astor and the St. Nicholas.
The streets leading off Broadway were often deep with mud,
and the sidewalks on Wall Street and Pine Street were a foot
higher than the road, a good protection for the pedestrian
against the wild drivers of the day.

By 1835 there were two hundred cabs in New York, which,
in proportion to the population, was twice as many as cruised
the streets of London. The law fixed the price: 37½ cents for
a trip less than a mile, 50 cents for a ride more than a mile.
Extra passengers were charged a quarter, and children from
two to fourteen got a 50 percent reduction. The flat rate to
Harlem with three hours of waiting time was $4, and the all-
day trip to King's Bridge cost $5. For attending a funeral
within the lamp-and-watch district—that is, where there were
streetlights and police—the flat fee was $2. But although all
these prices were set by law, hackies had the habit of asking

two and three times the legal fare. Almost everyone relied on
buses, which charged a uniform 12½ cents' fare inside the
city. To such far-flung roosts as Yorkville, the fare was 18¾
cents, and to Harlem it went up to 25 cents. As one English
visitor wrote in 1859, "No one thinks of taking a cab in New
York. Their fares are so exceedingly high, and you have to
put up with an amount of incivility from the driver unequalled
in any city in Europe." New York cab rates are no longer
considered high compared with other cities, but the heritage
of the cabbies' personality goes back over a hundred years.

Getting from New York to Philadelphia was, if not a major
odyssey, at least an involved excursion. One method was to
embark for the Jersey Coast by boat, then board a stagecoach
to Trenton and another boat from there to Philadelphia.
Charles Dickens, during his five months' tour of the United
States, made the trek with a train and two ferries. He noted
that it usually takes between five and six hours. Most travelers
were wearied by Philadelphia regularity. Dickens felt, after
walking about in it for an hour or two, that he "would have
given the world for a crooked street." No traveler failed to
make a pilgrimage to Philadelphia's most famous sight, which
was not the home of the Liberty Bell but the city waterworks.
Water was forced from the Schuylkill to a large reservoir, the
mechanism proving so fascinating in its day that special stages
were added to make the run out to Fairmount each evening
for those who wanted to see the phenomenon in action.

Baltimore fared rather better with early-nineteenth-cen-
tury voyagers. Mrs. Trollope, the English traveler, thought it
one of the handsomest cities in the Union to approach. She
remarked on "the noble column erected to the memory of
Washington and the Catholic Cathedral with its beautiful
dome seen at a great distance." As you enter the Baltimore
street," she wrote, "you feel that you are arrived in a handsome and populous city." Baltimore, known then as the city
of monuments, also had one of the nation's best inns—the City
Hotel, run by David Barnum and often called Barnum's.
Built in 1825, it stood six stories high, was filled with expensive fittings, was the *dernier cri* of the 1830s and 1840s and
lasted until 1889. Dickens, in 1842, called it, "the most com-

fortable of all the hotels of which I had any experience in the U.S. and they were not a few . . . where the English traveler will find curtains to his bed . . . and where he will likely to have enough water for washing himself, which is not at all a common case."

As early as 1816 Baltimore had a handsome theater with two rows of boxes set in a semicircle, an upper gallery, a pit on the ground floor, and Corinthian columns for decoration. It played to 500 persons and offered melodramas and such English fare as *She Stoops to Conquer.*

None of the niceties which caressed the traveler in the genteel preserves of Baltimore could, alas, be found in the nation's new capital at Washington. Visitors excoriated it, criticizing its architecture, its location, and its inaccessibility in a way that is remindful of the barbs one first heard about Brasilia, Brazil's daring capital hacked out of the bush a century and a half later. Wrote a Hungarian revolutionary who toured the nation with Kossuth, the Magyar patriot, in 1852: "Washington is like the eastern metropolis of a half-nomad nation, where the palaces of the king are surrounded by the temporary buildings of a people, held together only by the presence of the court."

The words of the Hungarian were like a love song compared to the appreciation left by Charles Dickens, who had seen the capital a decade earlier. "Washington," he declared, "is the headquarters of tobacco-tinctured saliva." Dickens' devastating critique went like this:

> Take the worst parts of the City Road and Pentonville, or the straggling outskirts of Paris, where the houses are the smallest. . . . Burn the whole town; build it up again in wood and plaster; widen it a little; throw in a part of St. John's Wood; put green blinds outside all the private houses, with a red curtain and a white one in every window; plough up all the roads, plant a great deal of coarse turf in every place where it ought *not* to be; erect three handsome buildings in stone and marbel, anywhere, but the more entirely out of everybody's way the better; call one the Post Office, one the Patent Office, and one the Treasury; make it scorching hot in the morning

and freezing cold in the afternoon [he was there in late winter] with an occasional tornado of wind and dust; leave a brick field without the bricks in all central places where a street may naturally be expected, and that's Washington. . . . It is sometimes called the City of Magnificent Distances but it might with greater propriety be termed the City of Magnificent Intentions; for it is only on taking a birds-eye view of it from the top of the Capitol that one can at all comprehend the vast designs of its projector, an aspiring Frenchman.

Pittsburgh's reputation for smoke and steel also had early beginnings. Dickens was less than enchanted by the smoky scene that greeted him on his arrival there. "We emerged," he wrote, "upon that ugly confusion of backs of buildings and crazy galleries and stairs which always abuts on water." But the situation of the city on the Allegheny River pleased his eye, and he was appreciative of the villas of "the wealthier citizens sprinkled about the high grounds of the neighborhood." An American traveler, Charles Fenno Hoffman, who journeyed west in 1833, found a sound on the streets of Pittsburgh that was unheard in any other town in the nation. "It is," he wrote, "the ceaseless din of the steam engines. Every mechanic of any pretention has one of these tremendous journeymen at work in his establishment." Pittsburgh, Hoffman decided, "is one of the most substantial and flourishing but least elegant cities on the continent."

Dauntless travelers who invaded the West in mid-century found Chicago's streets covered with boards to make what was called a "plank road." Water flowed under the boards and so did the sewage that emptied from the houses. St. Louis hotels were a maelstrom of activity with men crowding the lobbies on their way from Chicago and Cincinnati or just on their way there. Western hotels were filled with railroad builders, real-estate dealers, grain and lumber merchants and speculators. It was no place for an idler. "Porters, waiters, guest all are in quick motion; and one or the other is pretty sure to knock him over," one American traveler wrote. "In the great western hotels the tide of travel ebbs and flows twice in twenty-four hours. After nine in the morning rooms are easy to be had; after nine in the evening they can rarely be obtained for money and never for love." In St. Louis Dickens put up at the Planters House, which he said was "built like an English hospital, with long passages and bare walls and skylights above the room doors for free circulation of air." He found Cincinnati "a beautiful city, cheerful, thriving and animated . . . clean houses, well paved roads, and footways of bright tile." In Louisville he was at the Galt House, "A splendid hotel, and we were as handsomely lodged as though we had been in Paris."

Even as far west as St. Paul the explorer would find the first-class Fuller House, a five-story brick hotel that cost more than $100,000 when it was built in the 1850s. Of the southern cities, New Orleans was a nest of color not to be missed. Travelers coming in from the south across Lake Pontchartrain would transfer to a train six miles outside the city and ride into town on the rails and ties that had been laid atop the pilings. It was an hour's ride across swamplands, under cypresses dripping with Spanish moss, passing, now and then, a French tavern. The houses were built with high roofs and porte cocheres and the roads were lighted with lamps hung by ropes from tall posts. "We might indeed have fancied," mused one American, "that we were approaching Paris, but for the Negroes and mulattoes, and the large verandahs reminding us that the windows required protection from the sun's heat."

The St. Charles was New Orleans' best hotel before the Civil War. John Milton Mackie, a professor at Brown and author of a travel book called *From Cape Cod to Dixie*, wrote of seeing "gentlemen in purple pantaloons," and twice a day "a perfect flood of jewels, laces, painted silks, and muslins" coursing through the parlors. "It was," he wrote, "as if all the magnificence of Red River and Arkansas had emptied itself into the public rooms. And when the flood recedes one might be entertained with the sight of a couple of lovers on a sofa, sucking each an orange, while they estimate the value of their respective crops for the next season and speculate on the probable price of Negroes."

Even at the St. Louis, the best hotel after the St. Charles, the rooms were decorated in French fashion with muslin cur-

tains, scarlet draperies and drawing rooms in the mode of
Louis XIV.

If the cotton widows and the cotton girls of New Orleans
displayed their newly purchased finery in the parlors of the
St. Charles, along the eastern seaboard at Charleston the
southern ladies did their parading at the races. The February
meeting was well known throughout the East, famous for
beautiful women in their *grande toilette* and fine horses
groomed with equal care. If one was not promenading at the
races, then there was the green grass of the Battery with its
view of the bay and its ships coming and going. And if not
the Battery then the cemetery, a favored place in many American cities for a shaded walk, or so noted a Viennese concert
pianist who toured America in 1846.

In the soft climate of Charleston some traveled northerners
found reminiscences of Rome, and they came here to loll away
the winters on their spacious southern balconies. Others—they
were, of course, the most daring of the day—did indeed venture to Rome itself, to Switzerland, Paris and London. Although
the first passports, issued in 1796, had merely announced the
bearer as a citizen of the United States, the form was changed
in 1827 to include a description of the traveler.

Passports were issued free until July 1, 1862, when the
government decided, in order "to provide internal revenue to
support the government and to pay interest on the public
debt," to charge $3. In 1864 the fee was raised to $5, but in
1878 it was lowered to $1.

America got into the transatlantic steamship service with
the Black Ball Line, a packet company established in 1816.
Four ships, the *Amity*, the *Courier*, the *Pacific* and the *James
Monroe*, each about five hundred tons, sailed back and forth
between New York and Liverpool. The record for the eastward run was fifteen days and eighteen hours, but it was not
altogether unusual for the westbound passage, when the boat
was crowded with emigrants, to take forty days. Although the
ships provided the emigrant passengers with little more than
water and deck space (they were supposed to bring their own
food), the cabin passengers fared quite comfortably. The
dining saloon of the *Pacific* was 40 feet long and 14 feet wide,

with a long table running down the center and upholstered seats on either side. Seven private staterooms were laid out on either side of the dining saloon. The longboat, lashed to the deck, would serve as a pen for the sheep and pigs. Milking cows were kept in a barn over the main hatch and there was a coop for chickens. Thus America put to sea to explore the wonders of Europe.

Harriet Beecher Stowe, who sailed for the Old World in mid-century, found some beauty in the sea, but very little in the voyage itself. "In the first place," she wrote in a two-volume book published two years before *Uncle Tom's Cabin*, "it is a melancholy fact, but not the less true, that ship life is not at all fragrant; in short, particularly on a steamer, there is a most mournful combination of grease, steam, onions and dinners in general, either past, present, or to come which floating invisibly in the atmosphere, strongly predisposes to that disgust of existence, which makes every heaving billow, every white capped wave, the ship, the people, the sight, taste, sound and smell of everything a matter of inexpressible loathing."

Despite the poor press from people like Harriet Beecher Stowe, steamship travel on the transatlantic lanes attracted its adherents. By 1858 the famed *Great Eastern* was launched in England, a ship designed to carry more passengers than the *Queen Mary* and too big to have passed through today's Panama Canal. She arrived in New York with thirty-five paying passengers, including five newspapermen who paid first class fares of $125 each. She carried a crew of 418 and left a wake, as someone wrote, "like two bridle paths and a turnpike." When she docked in the North (Hudson) River, a great carnival spirit wafted up from the wharves where booths dispensed Great Eastern lemonade, and Great Eastern Oysters. Visiting farmers were stored in a makeshift waterfront inn called, naturally, the Great Eastern Hotel. The city went mad with excitement while the crew spent five days tidying up the ship. Then, at last, it was opened to visitors at $1 a head, a horrendously high charge for the time. After a week the price was dropped to 50 cents, and ultimately the Great Eastern racked up 143,764 admissions. Then she left on a two-day

excursion to Cape May, New Jersey. Two thousand passengers
at $10 a head signed on, and there began a wild seagoing orgy
of crap games and fistfights. There were not enough berths,
and passengers slept on the deck under a steady rainfall of
cinders from the five funnels. Food ran short and so did tem-
pers. When the *Great Eastern* arrived in New York the *Times*
wrote, "The Great Eastern has returned to the city and is
advertised to start immediately for Annapolis Roads. Don't go."

Ultimately the *Great Eastern* not only laid the Atlantic
cable but reduced the eastbound run to ten days. Although
she seemed always to be in trouble from the rigors of the sea
or the ineptitude of her managers, *Great Eastern* lasted for
forty-one years and remained the world's largest vessel until
the *Lusitania* was launched in 1906.

Over a hundred years ago American visitors on their way
abroad fretted about foreign customs inspectors. The inspec-
tions imposed by England in the 1850s, however, appeared to
have been as civilized in that nation as they are today. Mrs.
Stowe reported that "the customs officers, very gentlemanly
men, came on board; our luggage was all set out and passed
through a rapid examination which in many cases amounted
only to opening the trunk and shutting it and all was over.
The whole ceremony didn't take two hours."

American travelers noted, as early as 1850, that although
they could travel across their own country without being asked
for papers, they were stopped at every turn in Europe. At
each city the stage driver drew up at the gates while an official
demanded to see the passports, and if the party was merely
passing through town, they were stopped at both ends of the
metropolis. Hotelkeepers were required, as they still are today,
to report all arrivals to the police. Before moving on passports
had to be visaed by the minister of the next province or the
next kingdom, by the resident U.S. minister and finally by the
police.

A description of an arrival in Rome which appeared in the
Missouri Republican in 1850 gives a vivid idea of the security
regulations that existed:

We had hoped to have entered Rome by sunlight [but]

it was half-past ten at night when we arrived at the outer
gate of the Eternal City, which was shut, and called for
admittance. Safe within the walls and the gate locked, we
were next assailed by the police officers with their usual
bow-wow-ing, not a word of which we could understand
but passport, of which the Italians contrive to make four
syllables. These were all taken away and carried off and
receipts given for them; and then we were driven to the
custom house where our baggage was searched again, not
withstanding its recent overhauling at Civita Vecchia,
and before we reached our hotel it was past one o'clock
in the night.

Although, as one correspondent wrote, "We are sometimes
compelled to think that the chief end of man in Italy is to
attend his passport," the inconveniences visited upon travelers
from abroad did not seem in the least to dissuade them from
long safaris across Europe. "It is amazing to see the shoals of
people traveling this season on the continent for pleasure,"
wrote the *Missouri Republican's* correspondent in 1850. "Although the travel has been increasing for several years, it has
far exceeded this summer any previous one. . . . We were
repeatedly told that more Americans had visited the continent
this year than ever before. . . . We constantly found amusement as we rode or walked about in meeting travelers in
Switzerland all equipped in the same style, with traveling
bags or knapsacks slung over their shoulders, a staff in one
hand and an open guide book in the other, going on a pilgrimage among the Alps. . . ."
Equipment for launching a trip across Europe was outlined
in the guidebooks of the day. One of the most famous of the
early ones was *Information and Directions for Travellers on
the Continent*, by Mariana Starke, which was published in
London in 1824. In it Miss Starke made a list of the articles
which she considered useful to travelers in general and "needful to Invalids."

Leather sheets, made of sheepskin or doe-skin—pillows
—blankets—calico sheets—pillow cases—a musquite-net,

made of strong gauze, or very thin muslin—a traveling
chamber-lock—(these locks may always be met with in
London; and are easily fixed upon any door in less than
five minutes)—Bramahlocks for writing-desks and coach-
seats—a tinder-box and matches—a small lantern—towels,
table-clothes and napkins, strong but not fine—pistols—a
pocket-knife to eat with—table-knives—a carving-knife
and fork—a silver tea-pot—or a block-tin tea-kettle, tea-
pot, tea, and sugar-canister, the three last so made as to
fit into the kettle—pen-knives—Walkden's ink-powder—
pens—razors, straps, and hones—needles, thread, tape,
worsted, and pins—gauze-worsted stockings—flannel—
double-soled shoes and boots, and elastic soles; which are
particularly needful, in order to resist the chill of brick
and marble floors—clogs, called Paraboues; which are to
be purchased of the Patentee, Davis, Tottenham-Court-
Road, No. 229—warm pelisses, greatcoats and travellings-
caps—*The London and Edinburgh Dispensatory; or the
Universal Dispensatory*, by Reece—a thermometer—a med-
icine-chest, with scales, weights, an ounce, and half-ounce,
measure for liquids—a glass pestle and mortar—Shuttle-
worth's drop-measure, an article of great importance; as
the practice of administering active fluids by drops is
dangerously inaccurate—tooth and hair-brushes—portable
soup—Iceland moss—James's powder—bark—salvolatile—
aether—sulphuric acid—pure opium—liquid laudanum—
paregoric elixir—ipecacuanha—emetic tartar—prepared
calomel—diluted vitriolic acid—essential oil of lavender
—spirit of lavender—sweet spirit of nitre—antimonial
wine—super-carbonated kali—court-plaster and lint. A
strong English carriage, hung rather low, with well-
seasoned corded jack springs, iron axletrees, and sous-
soupentes of rope covered with leather—strong wheels
—anti-attrition grease—strong pole-pieces—a drag-chain,
with a very strong iron-shoe; and another drag made
of leather, with an iron hook—a box containing extra
linch pins, tools, nails, bolts for repairing, mounting
and dismounting a carriage—this box should be made
in the shape of a trunk, padlocked, and slung to the

hind-axletree—one well, if the carriage be crane-necked; two, if it be not—a sword-case—a very light imperial —two moderate-sized trunks, the larger to go before— a patent chain and padlock for every outside package— lamps, and a stock of candles fitted to them—a barouche seat, and a very light leather-hat-box, or a wicker basket with an oil-skin cover suspended under it. The bottom of the carriage should be pitched on the outside; the blinds should be made to bolt securely within side and the doors to lock. A second-hand carriage, in good condition, is preferable to a new one; and crane-necks are safer than single perches; though not necessary. Wheels made for traveling on the Continent should neither have patent-tire, nor patent-boxes; mail-coach, or common brass boxes, answer best. In those parts of Germany where the roads are bad, it is advisable to cord the wheels of traveling carriages; and the mode of doing this effectually is, to attach the cords to iron clamps fixed on the tire; after-ward fastening them round each nave. Every trunk ought to have a cradle; that is, some flat smooth pieces of oak, in length the same as the inside of the trunk, about two inches and a half wide, nearly half an inch thick, and crossed-barred by, and quilted into, the kind of material used for saddle girths; a distance of three inches being left between each piece of wood. This cradle should be strapped very tight upon the top of the trunk (after it has been packed) by means of straps and buckles fastened to its bottom; and thus the contents can never be moved, by jolts, from the situation in which they were originally placed. Every trunk should have an outside-cover of strong sail-cloth painted.

Persons who travel with their own sheets, pillows, and blankets, should double them up of a convenient size, and then place them in their carriage, by way of cushions, making a leather-sheet the envelope.

Ten drops of essential oil of lavender, distributed about a bed, will drive away either bugs or fleas; and five drops of sulphuric acid, put into a large decanter of bad water, will make the noxious particles deposit themselves at the bottom, and render the water wholesome; twenty drops of diluted vitriolic acid will produce the same effect.

Persons who wish to preserve health, during a long journey, should avoid sitting many hours together in a carriage; by alighting and walking on, while their horses are changed, provided they travel post; and by walking up all the ascents, provided they travel en voiturier; and persons who get wetted through, should take off their clothes as soon as possible, rub themselves with Eau de Cologne, and then put on dry warm linen, scented with Hungary water.

I will now close this subject by observing that Travellers should never fail, before they enter an Inn upon the Continent, to make a strict bargain with the Landlord, relative to their expenses; and bargains of every description should be made in the currency of the country.

Letters sent back to the *Missouri Republican* noted that on the continent one is "not charged by the day for three meals whether you eat them or not, but so many florins, or francs, or batzens or lires or pauls or carlins according to the country you are in, for your room and for as many meals as you take. . . ." This rather sensible system became known in travel circles, as indeed it still is, as the European plan. The American plan quoted prices which included three meals.

European hotels, which are still disposed to exact extra charges for winter heating and summer air conditioning, charged extra in these days for candles and fire. If two flaming wax candles were brought to a guest's room he would find a 20-cent charge on his bill. One gentleman traveler found he had in a few weeks' time accrued a charge of $25 for candles. Wise travelers were advised to bring their own candles with them. And to ask the price of a room as soon as it is shown. "The people here," the *Missouri Republican*'s traveling correspondent wrote, "seem to think that nothing is too good for the English and Americans, and moreover, that they are made of silver and gold." Just how much silver and gold was needed is shown by the report of a mid-century traveler who spent four weeks in England, two weeks in France, three weeks in Switzerland, one week on the Rhine, and one week in Belgium. He spent $600, a figure which included four months of travel including staterooms on the packet to England and the steamship back to the United States.

The mid-century traveler from America could spend four weeks in Europe on a budget of $600 including staterooms across the Atlantic. He journeyed with Murray's Guide, stopped at Meurice's Hotel or at the Hôtel de Louvre in Paris and at the Baur au Lac in Zurich, a fine hotel which offered black tea and *Galagnani's* English newspaper, both in short supply east of Paris. Most Swiss hotels served a *table d'hôte* dinner, at twelve-thirty for the Germans and at six for the English. Gold watches were on sale in Geneva for $40 and a visitor from New Orleans noted that diamonds and precious stones were about half of what they cost back home. Interlaken was even then a favored watering place frequented by the Empress Dowager of Russia and hordes of English travelers. "Lord Snob and Lady Upstart can be seen every evening riding out in fine carriages and liveried servants." But Switzerland was crowded with Americans, so many, in fact, that one traveler wrote that "like the locusts of Egypt they are infesting the land." What were all these Americans doing abroad in these pioneering years of European travel? Well, they were: carving their names on John Calvin's pulpit in Geneva; signing in on the altar of Notre Dame in Paris; visiting the tomb of Lafayette, in 1850 still a big hero to Americans; ruing the lack of ice in Liverpool and the impossibility of getting a mint julep; looking wide-eyed into the leakproof tunnel of the Thames, a curiosity of the city; stepping off the distance William Tell shot the arrow in Fluelen (130 yards); gambling in the ornate saloons of Baden-Baden; threading through the canals of Venice in gondolas; and listening to Johann Strauss the younger play with his band every Sunday at a fashionable retreat in the outskirts of Vienna.

They were the good years, but as always, they were not to go on forever. Back home Newport, since its first days a favorite with southern families, was now attracting New York patrons as well. Dancing, flirting, riding and bathing were the excitements of the time. At Newport the ladies, dressed in pantalettes that reached to the ankles and red frocks that reached to the wrists, bathed in company with the men who guided their partners through the surf "as if," one observer noted, "they were dancing water quadrilles." Down in the South, at Old Point Comfort, the belles were not quite so daring, confining themselves to segregated stockades built into the water. Atlantic City began its emergence in the late 1850s, and so for that matter did Coney Island, the beach front in Brooklyn, famed for its roasted clams and toasted bathers.

Before the Revolution the northern resorts had provided a meeting place where colonists from all parts of the East Coast came together and discovered their common problems and common animosities. A century later it was different. The meetings between sectionalists in the northern resorts even served to magnify the differences. Nor did it seem to help that Senator Jefferson Davis of Mississippi as late as 1858 was in a party of vacationing southerners who always spent the summer weeks in Maine. Or that a proslavery feeling permeated the hotels and even the churches of Newport, which had always played host to so many Southerners.

Not all resorts were as friendly to the Confederate cause. Some southerners came home from vacations in the North smarting from northern jibes. "We are treated worse in the north than if we were foreign enemies," the Richmond *Whig* reported. "Let us with one accord stay home and spend among our own people." Southern stumpers snatched at the cry, railed at the northern resorts, demanding a boycott. Dixie began to look more closely at resorts of her own. The planters began to build summer villas at Virginia's mineral springs, and White Sulphur surged with new popularity. A stock company was formed to back southern resorts. One of its first projects was a Southern Saratoga, which they hoped to create at Montgomery, Virginia, near the Tennessee border. Summer homes sprang up along the Gulf Coast of Mississippi, among them the mansion of Jefferson Davis, which is still standing. So the virus of sectionalism infected even the circles of pleasures and holidays until at last on an April day in 1861 Confederate forces fired on Fort Sumter at the harbor entrance to the enchanting city of Charleston, where so many northerners had in other days come to loll away the winters, watching the belles parade their *grande toilette* at the February races and likening the climate to Rome. The war was on and pleasure was done—at least until Appomattox, just two days less than four heartbreaking years away.

III | *Vagabonding Victorians (1865–1899)*

By THE TIME the Civil War had ground its sad and wearisome way to an end America was more than ready for a fresh look at living. Pleasure became the byword and money was the key. With no income tax and the country striding westward clear to California, the opportunities for making money seemed limitless. There were also some prime opportunities for spending it. One could, with means, ride the rails in relative comfort, and the tracks were spreading farther and farther across the United States. Barely twenty-five years before, the Pacific Coast had been no closer to the centers of Eastern population than Bombay. But on May 10, 1869, the Union Pacific, coming from the east, and the Central Pacific spluttering in from the west, kissed cowcatchers at the metropolis known as Promontory Point, Utah.

The first grand curlicued vestiges of the approaching Victorian era crept into the decor of the railway carriages. And as their routes spread in new directions, the railroads anchored each important terminus with a great armory for the transfer of passengers, some sufficiently grand to have pleased a French Louis and perhaps to have received a mummified pharaoh in transit to the Egyptian hereafter. The nation's riverboats, and ultimately the transatlantic liners, followed this ornate lead. The hotels that were destined to rise in this fussy era would prove a fantasy cooked of yeast, gingerbread, frills and grand plan—a recipe that was to produce, among other elaborate caserns, the Ponce de Leon in St. Augustine, the Lafayette in

Minnesota and the del Coronado far out in San Diego.

Saratoga, which sagged a bit with the early loss of its gay southern clientele in the years immediately following the war, made a rapid recovery. Summer vacations, which had once been a dalliance reserved for the well-to-do, now became a custom of the white-collar class. So many people flocked to Newport, to Long Branch and the familiar watering places that the *haut monde*, a bit miffed to find such a conglomerate lot inhabiting the hotels it had once counted as its private preserves, began to build summer cottages of its own. At Newport any shelter designed for seasonal use was called a cottage, even if it cost $100,000 or more. For those who in this bountiful time had not yet accumulated that much cash, there was summer camping, a sportive recreation that was just beginning to evoke a certain interest. And among the leisure class there were early rustlings of yet a new form of indulgence: winter vacations in Florida. Jacksonville, which winter sun-seekers these days would as soon visit as Sioux City, was doubling its population every time January rolled around.

It was the railroads and their newly designed comforts that made possible the emerging period of domestic travel. In a summer story in 1866, the New York *Times* reported that Pullman's sleeping car *Omaha* measured 70 feet long and ran on 16 wheels. The interior was 10 feet wide and 10 feet high and "affords cool, comfortable and clean couches for 64 persons." The interiors were paneled in black walnut, carved of course, the seats covered with velvet, the corridors paved with carpet. Damask hung at the windows, fragrant bouquets were suspended from the ceiling. It all looked, said the *Times*, more like "an elegantly furnished parlor than the interior of a railroad car."

Bookings were so good on the first sleeping car that George M. Pullman formed a company to build twenty of them at $20,000 each. Still there is evidence that not all postwar train travel was cushioned in Victorian frippery. Only a few months after it had heralded the arrival of the Pullman sleeping car the New York *Times* leveled an editorial fusillade at some shortcomings in rail travel. It scored the "shameful treatment of luggage," noticed that boxes were being broken and "trunks smashed without the slightest compunction by the roughness of the employees." Said the *Times*, "This has become a crying evil, and nowhere is it more flagrant than at Saratoga." The paper also noted, while it was on the subject, that the time had come for some enterprising companies to run extra-priced cars more elegantly fitted up "for ladies or family parties or those who do not wish to be annoyed by tobacco spit or drunken or dirty company."

Whether it was because of the editorial pleadings or because the public clamored for more luxury isn't clear. But soon a splendor rode the rails such as has not been seen to this sybaritic day. After the sleeping car came the palace car with its own divided "drawing rooms," then the "hotel car" with its own kitchen and portable tables that could be placed between the seats. The first one, the *President*, rolled over the tracks of the Great Western Railway of Canada in 1867.

Special effects turned from the luxurious to the bizarre. One palace car company dressed its conductors in red lined capes. The New York and New England put on a *White Train* with the engine, coal car and carriages all painted white, white silk curtains at the window, velvet carpets on the floor. The engineer was done up in white overalls, white gloves and white cap. Lucius Beebe and Charles Clegg, the famed railroad chroniclers, insist the coat was whitewashed before every run.

The Chicago and Alton offered not merely grand menus and grand decor. They dispensed culture. While most railroad timetables contained train schedules and advertisements for hotels and resorts, the Chicago and Alton undertook to run the complete works of Robert Browning serially. The series began in 1872 with *Pauline* and went monthly without pause except for one summer sequence when the railroad tempted its passengers with views of the hotels and the landscapes of Wisconsin and Minnesota; but in September culture was back riding the rails.

In the ornate, extra-fare trains the menu matched the decor. Champagne and claret were served for breakfast, buffalo tongue and bear steaks appeared on the menus in the West, scrod and oysters in the East, terrapin, Georgia peaches and Florida melons in the South, sand dabs and California figs in the West. The menu on the Chicago Burlington and Quincy in the 1870s was like a lexicon compiled by Escoffier. Beginning with saddle rock oysters on shell, it cruised through four kinds of soup, rolled past pheasant larded with truffles, sweetbreads, turkey with oyster sauce, and mutton kidneys, sped onward to quail, canvasback duck, English snipe and broiled woodcock on toast. It steamed to a grand finale with English plum pudding, fruitcake, French kisses, ice cream, oranges, nuts and grapes. All meals: 75 cents.

Dining cars carrying such larders were only hooked onto extra-fare trains. The ordinary traveler was obliged to grab a snack at the trackside cafés where the trains stopped for twenty minutes. Candy butchers were put aboard both the hotel trains and the immigrant cars, and they peddled sandwiches, candy, pillows, soap, towels, cigars and papers, magazines, dime novels, and genuine Indian souvenirs. There was some improvement in trackside dining when Fred Harvey strung a chain of restaurants along the nation's rails. When he died in 1901 his company had accumulated fifteen hotels, forty-seven restaurants and fifteen dining cars. A company that still bears his name continues to operate in the West and its newest facilities lie alongside superhighways.

Resorts soon realized that to survive they would have to have a rail line to bring in the customers. Soon after the war a railroad committee met at White Sulphur Springs to establish a line that would eventually become the Chesapeake and Ohio. But the famed Old White didn't wait for the train. It reopened in 1866 with a racecourse and new artificial lake. The old Treadmill custom was immediately revived and guests once more tramped through the grand salons greeting old friends, looking for likely new ones and poring over the hotel register, which was displayed in the parlor. The music from the ballroom would at length announce the hour for the White Sulphur Riley, a dance that, while related to the minuet, was exclusive to the Old White.

General Robert E. Lee returned to the Old White in 1867, and although the exact hour of his arrival was not known, over a thousand people were waiting for him. When the glass doors opened and he strode into the dining room, all the diners rose.

Later that evening, when he had retired to his cottage, which was called Baltimore G., Professor Rosenberger and his famed Baltimore band assembled on the lawn to serenade him.

General Beauregard and ex-President Jefferson Davis were among the many other Confederate luminaries who came that year. A ball was arranged in honor of General Lee and a few days later he held a reception of his own with thousands of his old soldiers filing through a drawing room to shake his hand. Lee came again the next year and again in 1869, the year before he died. In midsummer seven years after his death, the Old White staged the Robert E. Lee Monument Ball to raise funds for a statue of him to be erected in Capitol Square at Richmond. The event seemed to mark the return, at last, to the great social days that flourished before Sumter. Said the Richmond *Whig* somewhat breathlessly:

The present season at this famed watering-place is the most brilliant known since the war. Crowds continue to come in from East and West. The Lee Monument ball, which took place last night, fulfilled the most sanguine expectations of its originators. The ballroom was not large enough to hold the brilliant concourse of fair ladies and handsome men which gathered on the occasion. Accordingly the spacious dining-room, 320 feet in length and 140 feet wide, was secured.

One thousand pretty ladies and good-looking gents skipping the light fantastic beneath the glare of two locomotive headlights and a hundred lamps, was a decidedly attractive spectacle to look upon and the well-glazed floor presented a fine field for the dancer, whether he might be the amateur who would fain hobble about fearful of mistakes or the practiced devotee of this noble art of skipping over trouble and time.

The ball was a grand success. The ladies' toilets were elegant and rich, while the male heirs of Adam's real estate were gotten up regardless of expense.

A good sum will be realized for the Lee monument. This ball has shown to the politicians here that there is "no North, no South." Now the ex-Confederate General

danced with the belle of New York, while the brave
Union Captain was foremost in the dance with his fair
Southern partner, and a forest of Colonels (Virginians as
well as Georgians) shaded the scene and gently breezed
each passing damsel. The ladies and gentlemen from the
North have worked nobly for the Lee monument. No
further trouble need be apprehended. The country is
safe. The angel of peace or a near relative hovers around.
All is over. At the Lee Monument ball the differences
between North and South were healed and toed, and we
can now present to the gaze of shaky monarchies a whole
shoe with here and there a patch, it is true, but strong as
when it was new.

The ball has been described by many newspaper correspondents, and they have especially given prominence
to the ladies' toilets, as has been their custom on like
occasions. Each movement of the fair creatures was the
poetry of motion; their eyes shone like countless stars; in
silks and satins they were gorgeously grand; lips rivaled
the coral; teeth were as pearls; hair glossy and glorious;
diamonds gleamed and twinkled above coronets of splendid tresses. All this has been told with all the wealth of
language that the newspaper correspondent could invent,
borrow or steal. That is all. Not a word about the men
who invested one hundred dollars for a broadcloth suit,
five for white vest, three for kid gloves, twenty-five cents
for blackening shoes, two dollars for shave, shampoo, and
bath, dollar for necktie, two for silk handkerchief, five for
ball ticket and ice-cream—total one hundred and eighteen
dollars.

Oddly enough, the very same summer, the New York *Daily
Tribune* suggested that some of the horde that was surging
through the northern resorts be diverted to the gentle dells
of the South. It might seem a reversal to suggest that northerners travel south in summer, but it was noted that the
wealthier classes of Virginia were already at their summer spas
just as they were before the war. Not only would a southern
vacation prove pleasant to the northerner, the paper said, but

it might also prove helpful to the younger generation of southerners who had shown themselves too ready to fall back on "the fortune lost in the war" as "sufficient for present and future glory." Meetings with northerners might give them fresh ideas.

Besides, said the *Tribune*, "The whole mountain region from Upper Virginia to South Carolina offers temptations such as are not to be found in the Adirondacks to sportsmen of every kind." It reminded New Yorkers that there are "quiet little watering places . . . where the Northerners receive an exceptionally cordial welcome. Board at these places . . . range[s] in price at from $8 to $10 a week. There are quaint little inns, too, in villages lying literally above the clouds, where the same kind of board costs but from $3 to $5 per week; where bear steaks and venison collops appear for breakfast."

The only drawback, said the *Tribune*, were the "extortionate charges" made by every southern railway wayside hotel. They are usually three times as high as the rates in the North because the passengers are few and the hotels "must make of them what they can." It was a suicidal policy, the *Trib* rued, as the hotels would shortly know from sad experience.

In the fashionable East the resorts famed before the war quickly regained their stature. President Grant lent new style to Long Branch on the Jersey shore, to which he retreated during the warmest months of summer. Following the vogue he built not one but two summer cottages, spent much of his time in long carriage drives about the country. Newport's fanciest cottage, Château-sur-Mer, owned by George Peabody Wetmore, cost an estimated million dollars. A rather unusual effect was created in the Wetmore dining room, where the fine wood walls were carved in fruits and flowers, giving the effect of an ornate frame for a large painting which covered the ceiling. So one didn't, as in an ordinary house, sit at a table and look *across* at a masterful painting, one dined under it. Pierre Lorillard built a similar showplace, Mr. Charles H. Russell of New York put up an Italian villa, and Fairman Rogers of Philadelphia made the first use of picture windows. Visitors

entered Mrs. Loring Andrews' cottage through an entrance
hall that was twenty-four feet square and seventeen feet high.
J. P. Kernochan of New York incorporated a private ballroom
with a vaulted ceiling. Mrs. Paran Stevens of New York installed stained-glass windows. H. G. Marquand's house was
furnished with so many oddments that a snappy writer of the
day suggested that it be called Bric-a-Brac Hall.

Although fabled mansions sprouted in Newport like summer mint, all of them, of course, were called cottages; there
seemed no similar effort to create luxury hotels. The community had to get by with the Ocean House, which had five
hundred adequate rooms, none of which, despite the hotel's
name, faced the ocean. Smaller and perhaps more quiet was
Acquidneck House, which wasn't on the ocean either. Both
charged $5 a day.

At Newport one followed a social regimen, bouncing from
breakfast to band concert to beach. Then joining shopping
excursions, promenades and back in time to dress for dinner.
It was quieter over on Block Island, where, as a guidebook
of the day put it, "elaborate toilets are the exception, not the
rule." It was cheaper, too. Ocean View House cost $3.50 to
$4 a day, with a liberal reduction at the first and last of the
season. It had room for six hundred, every room and every
cottage lighted by gas and connected to the office by an improved system of electric bells and a device known as Holtzer's
Patent Indicator. Quimby's orchestra, equipped with both
strings and brass, played for the circumspect celebrants.

Choosing the proper summer resort became an annual and
a seasonal problem. The New York *Daily Tribune*, as early as
April one year, cautioned its readers to make their selections
with care. "It is notorious," the *Tribune* said, "that few places
of summer resort are in every respect desirable. The water
supply is oftentimes liable to be contaminated and the arrangements for disposing of sewage are of the queerest sort."
It is noted that the outbreaks of typhoid in Brooklyn, a resort
of consequence, had in some cases been traced to exposure
during the summer season.

The papers also counseled their readers on what to wear.
The New York *Times* reported, at the outset of the 1890 travel

season, that, "As large and varied an outfit of equipments is now required for the fashionable traveler as was once demanded for the pioneer crossing the plains or the soldier going to the front. . . ." Traveling cloaks with a monk's hood and a cord girdle were appropriate for doing the monasteries and cathedrals. Gloria silk was the season's favorite material. Steamer trunks were down to $2.25, although you could spend up to $25 for one in solid sole leather. A drinking cup for water and a tourist glass were judged indispensable items for a trip.

Early in June of 1888 the New York *Daily Tribune* printed a list of useful items of clothing for the gentleman en route to spend the season at Newport, Bar Harbor, Saratoga or Lenox. A $1500 bill for clothing might indeed prove staggering, the *Tribune* admitted, but a prudent man of fashion could still keep the bill as low as $900. Here is the *Tribune's* suggested wardrobe:

1 lightweight dress suit	$110
1 black double-breasted frock coat & suit	100
1 gray frock coat & suit	100
1 3-button cutaway coat & suit	85
1 2-button cutaway coat & suit	85
2 sack coats & suits	150
1 covert coat	65
1 light overcoat	75
4 extra pairs of trousers	70
6 fancy waistcoats	60
1 tennis suit	40
1 polo suit	45
1 pair riding breeches	25
1 pair top boots for riding	15
1 silk hat	8
tennis & polo caps & 2 straw hats	10
3 pair dress shoes	45
3 pair calfskin shoes	40
tennis, polo & knockabout shoes	15
2 doz. linen shirts, dress & plain	100
neckties, collars & cuffs etc.	40

½ doz. flannel shirts	90 (40)
2 flannel suits for rowing & yachting	90
underwear, hose, etc.	150

The newspapers also ran pages of readers that were little more than glorified classified ads which told, albeit succinctly and without display, the glories to be expected on the premises. The Poland Spring Hotel in Maine advertised open fires and steam heat, as well as first-class accommodations "in connection with the 'Far Famed' Poland Water, Nature's Great Remedy, Greatest Medicinal Water in the World, Cures all Diseases originating from Dyspepsia, Kidney and Liver Complaints, and all diseases of the urinary organs."

The Long Point Hotel on Seneca Lake, New York, wished it known that it had "no mosquitoes, no malaria; bowling; archery; tennis; boating; bathing and fishing; good livery; a first-class orchestra; weekly hops. Table and service unsurpassed; transient, $2 per day and upward. Special rates for families."

Health was frequently an important consideration, possibly an evolution from the colonial days, when a trip to the spa in search of a cure was in reality an excuse for a vacation. A typical ad designed to tempt those seeking a well-equipped, well-rounded resort of the day read:

DR. STRONG'S SANITARIUM

Saratoga Springs, N.Y.

A popular resort for health, change, rest or recreation all the year. Elevator, electric bells, steam, open fireplaces, sun parlor, and promenade on the roof. Croquet, lawn tennis, Ec. Massage. Turkish, Russian, Roman, Electro-Thermal, all baths and all remedial appliances.

The *Tribune* seemed so excited by the number of advertisers who sought exposure in its columns that it ran an editorial drawing attention to its full page of ads, the result of "an imperious demand for space in which to set forth the advantages and disadvantages of the country." As the season rolled on the papers printed social notes from the resorts, the *Tribune's* frequently running under the standing headline, "Summer Leisure."

Leading the page, one August Saturday, was an account "from an occasional correspondent of the *Tribune*," writing from Bar Harbor, Maine, already a resort of consequence. Occasional Correspondent had taken a week to grope his way northward in the fog, but the sun had finally blossomed, inspiring him to quote from Homer about the beauties of the far Northeast. Since fog shrouded the mornings, Occasional Correspondent reported, "nobody thinks it worthwhile to be about before nine or ten." Such "late morning hours," as he called them, were not all that confounded the reporter, who was also rather startled by the dress of the vacationists, the civility of the staff, even the variety of the water craft. Wrote he:

> I, rising with the fog, look about me and see certainly not what I came forth into the wilderness to find. I behold men clothed in soft raiment or gorgeously gotten up in bed-ticking of ruddy stripe for the tennis field; women clad in more brilliant and varied apparel than at home flash upon my vision, driving by in vehicles as ugly and as stylish as the latest English importations on Fifth Avenue.

While drawing many vacationists from New York, Philadelphia, Washington and Baltimore, Bar Harbor had not as yet attracted "the great vulgar crowd," as the *Tribune's* man put it. It was far from the big cities, the beaches were few and the private holdings extensive and limiting. Bar Harbor in those days had one hotel, Rodick's, named after its builder, an informal place, despite its size, compared with the arch formality that starched the regimen at Saratoga and Long Branch. Although it had five hundred rooms and claimed to be the largest hotel in New England, it couldn't brag of a single private bath. But it often attracted crowds of three thousand to its dances, which were held twice a week; and if you didn't meet anyone there, well, there were always the promenades along the five hundred feet of porches and strolls through the great lobby which society referred to, unabashedly, as the Fish Pond.

The pastimes were otherwise tame and rather healthy at Bar Harbor, including, as they did, rock hunts along the shore, walks through the trails that wound across Mount Desert, and boating. "Everything is to be found here from the catamaran and the steam yacht down to the birch bark canoe," Occasional Correspondent reported. The boats, while reasonable to rent, were not as cheap as they were at Lake George. Canoes could be custom tailored and made right on the spot. "Ladies are fond of them," O.C. wrote, "perhaps because they can arrange picturesque tableaux, seated in the bows under gay parasols, while the boats glide hither and thither with an easy effortless motion."

At Newport the season was so good that hotels were forced to put up cots, and finally to turn guests away. The first fox hunt of the season was a success, Cornelius Vanderbilt made the social notes with a small dinner at Wales cottage, and the Stanley Mortimers had a hunt breakfast at Bryers. Up at Saratoga more than a thousand guests were stuffed into the United States Hotel. William H. Vanderbilt, who was staying there, took a party to a matinee performance at Rochester and was back in the hotel that night. A moonlight coaching club drove to the lake for a sail on a steam yacht, and the American Dental Convention was about to open its convention. Long Branch was jammed, Alstrom's Band had been engaged for concerts twice a week on the pier, and "pyrotechnic displays" were also announced for the same nights. The Atlantic Hotel, the Ocean, the Mansion House and the United States sent long lists of their arriving guests which were dutifully printed by the New York papers.

Asbury Park had emerged as a "unique city by the sea," its summer population swelling to twenty thousand, its avenues lined with brightly painted cottages, each with a broad piazza, and a private little nook just large enough for two. From the depot, stages ran all over the city charging five and ten cents a ride. The ocean walk was judged one of the best in the country. Coleman House opened a music hall which would seat a thousand people and had a stage for theatricals. Although it was used mornings and afternoons for roller skating, the floor was smooth for dancing, and so it was in use day and night. There was a billiard room and bowling alley on the lower floor, but, as the correspondent noted, it had one drawback. "The only thing to be obtained in the shape of a drink is soda water or lemonade." Undaunted, the guests swamped the reservations desk at the Coleman House and cots had to be put up in the parlors and reading rooms.

And the rush was on in the Catskill hotels which were not yet known as the Borscht Circuit. "The afternoon scenes on the dock at Kingston and at Catskill Point were full of bustle and excitement," the *Tribune's* local man telegraphed one weekend in the summer of 1881. "Everybody rushes for a seat in cars or stages. Trunks and boxes are sent crashing through the air regardless of results. And at Catskill, where 150 or more stages, hacks and wagons are in waiting, each with their noisy runners and hotel agents, the visitor has a sorry time indeed." The July season was better than ever, August prospects were "brilliant," and the Kaaterskill Hotel, full all week with notables from Washington, announced it would add a 300-foot extension in the fall. There was even a rumor that the owner was going to buy one of the buildings from the Philadelphia centennial, float it north by canal barge, and plant it atop a mountain as a resort.

Cape May, on the Jersey shore, which had suffered from a fire, never seemed quite to recover. It also had other problems, on which the *Tribune* commented quite candidly in its issue of July 29, 1881:

> . . . the war on Hebrews is maintained here. Last year Congress Hall was full of the Semitic race, and the "lineal descendants" of Lord Baltimore could not bring themselves to lodge under the same roof as the Hebrews; so the great piazzas and "the best piece of beach in the world" were given up to the Beaconfields and the Rothchilds. The proprietors seemed to enjoy this state of affairs no better than did the only people in the country who are unable to pronounce properly the place where they live. This year, therefore, Congress Hall is forbidden ground

to Hebrews. There has been no proclamation of their exclusion. It has not been advertised. There is great ability displayed, however, in convincing unwelcome guests that they will not be satisfied with the high prices and the worst rooms of Congress Hall, and they wander over to the Stockton.

The same era is responsible for one celebrated story about hotel discrimination which occurred at the famed Grand Union Hotel in Saratoga, which was owned by A. T. Stewart, proprietor as well of a famed department store in New York. Joseph Seligman, the banker, had engaged a suite of rooms at the Grand Union for himself and his family for many years. When, in the summer of 1877, the Stewarts decided that Jewish guests would henceforth be excluded, Seligman, in some disbelief, was turned away at the door. Since discrimination against Jews had been relatively a minor matter before the Civil War, particularly in social circles, the incident at the spa created excited ramifications that resulted in a Jewish boycott of Stewart's department store. Stewart was so badly hurt that he sold his store to John Wanamaker.

No such social difficulties seemed to face the new resort on the Jersey shore which was called Atlantic City. "Nowhere, except at Coney Island," said the *Daily Tribune* in 1883, "is there such a free and easy democratic resort as this. Everyone meets here on equal terms. Cabinet ministers and clam catchers wallow in the sand and surf side by side. Members of Congress sit at table alongside charcoal burners . . . children of Philadelphia millionaires and children of Absecon oystermen ride together on the merry-go-rounds. . . ."

There were only seven houses in Atlantic City when the railroad arrived in the 1850s. And when the boardwalk was first laid in 1870 there were about two thousand people living in town. The boardwalk, lighted at night and called by some "the most fascinating boulevard in the world," ran over soft sand, mosquito marsh, and a scattering of bathhouses. Brighton Cottage, which opened in April of 1876, announced that it would take guests all year round, thus making Atlantic City a winter resort, too. Still, it had its problems and the newspapers carped continually at the antique system of carting garbage away and dumping it, at the number of Philadelphia crooks who inhabited the place, at the poor security at the bathhouses, which were forever being robbed, and at the inadequate police system which removed officers from the beach area after eleven at night. Nonetheless, Atlantic City was considered a great place for children, and before the century was over there were $10 million invested in hotels and boardinghouses. *The Daily Union History of Atlantic City and County*, which came out in 1900, permitted itself a statement of some historical import on the subject which deserves to be handed down through the generations. "While Atlantic City," it said, "may not have palatial hotels to compare with the Waldorf-Astoria, New York; the Ponce de Leon, St. Augustine; the Palace Hotel, San Francisco; the Great Northern or the Auditorium, Chicago; Brown's Palace, Denver; the del Monte of Monterey or the Del Coronado, Santiago [sic], California, the same may be said of Philadelphia."

A number of resorts, in New York particularly, in addition to providing a holiday, insisted, at the same time, on improving the mind and the body. Most famous of all the cerebral spas was the celebrated station at Chautauqua in western New York State. Here among the leafy trees, by the shores of the lake, courses were given in biblical history, music classes resounded through the dells. Clusters of people were brought together for denominational prayer meetings, for lessons in Greek and Hebrew, for Latin, French and German. Even the children were marshaled and trouped off for Sunday school instruction and kindergarten exercises. From fifteen hundred to two thousand persons, from all parts of the country, might elbow into the assemblies. Some were day trippers who came with picnic baskets, but many stayed at the handsome cottages festooned with climbing vines, rimmed with flowerbeds and broad piazzas. In 1880 the Palace Hotel opened a new ten-thousand-dollar wing honeycombed with large, airy rooms, each with gaslight and running water, a big improvement over the older part of the house, where the rooms admittedly provided shelter only. The dining room was in reality a tent, but the boarded sides could be opened on fine days to a view of the lake. Musical entertainment scarcely got beyond revival and hymn singing, but the evening programs frequently drew an audience of ten thousand.

Inner solace and an improved constitution were the offerings at New York's Richfield Springs. "The air at this spa," a correspondent wrote in 1880, "is something wonderful; it quiets the nerves, induces sound sound sleep, arouses dormant energies, and incites people to great undertaking, political and otherwise, fits them to resume care and business with renewed vigor, making molehills of responsibilities that seemed mountains before." Not only were the mineral waters considered health-giving to ingest, they also chased the bugs, making them as deserving of canonization, one visitor said, as Patrick of Ireland. Lawn tennis was the popular game and fans the season's fashion eccentricity, much larger and brighter than last year. Belles are inclined to peep over them like a Cuban girl over her jalousie, a visitor wrote, the affectation perhaps deriving from the large number of Spanish Cuban visitors who came to the place.

In the era in which Chautauqua blossomed, Niagara Falls, just to the north, which had glazed the eyes of Charles Dickens forty years before, by now was in a sad state of disrepair. *Harper's Magazine* published a sharp rebuke in 1882. "More than once," it wrote, "we have spoken of the fatal injury done to the State of New York, and to the national character itself, by the desecration of Niagara Falls." It cited the falling off of visitors, the "complete vulgarization of the approaches" of the petty annoyances, the Indian shops, the hackmen, the unsightly structures that had arisen along the banks. "Let New York spare herself the shame of the practical obliteration of Niagara Falls."

But Americans had no need of Niagara; other places beckoned. They swarmed over the Thousand Islands, about three hundred miles north of New York City, and despite the inaccessibility, they were busy hammering together summer houses all over this newfound watery dell. Passengers who were landed at Clayton by the Utica and Black River Railroad had been on the road for fourteen hours. Those who were not staying at Clayton, which had two summer hotels, took a steamer for Alexandria Bay, where the action was. Thousand Islands House and the Crossman were the best hotels, and as Dr. J. G. Holland, editor of *Scribner's*, wrote in a long rambling report for the New York *Daily Tribune*, "They do not equal the Windsor or the Brevoort but they are luxury itself when compared with anything I have ever found on the coast of Maine." The twilights were so long up among the islands that Holland said he could almost read a newspaper at nine o'clock. Many wealthy visitors who had bought islands and built homes were now busy buying yachts. "This is the paradise of steam yachts," Dr. Holland wrote, and ". . . the lively chatter of their 'direct exhaust' fills the air."

The Adirondack Mountains of New York State were a new discovery, too, and much less civilized for summer living than the Thousand Islands. Those who ride through these mountains now and find their bumpers wired with an advertising sign printed in bold colors may find it hard to imagine that until the early 1870s, the Adirondacks were largely a trackless range dotted with camps operated by Indian guides. Blue Mountain Lake Hotel, a rather primitive log structure, opened in 1875 claiming to accommodate forty guests. Probably it could take fewer, but it was clean and well run, and the patronage faithful, even though on the stage line from the depot the road was so rough that the women had to be strapped to their seats. In the early years of the hotel the fifteen-mile journey took four hours.

The very remoteness of the Adirondacks was, in the late 1800s anyway, a great part of their charm. When the Prospect House was built in 1882 with porches, or piazzas, as they preferred to call them, running in 20-foot avenues for 370 feet, the place was dubbed the Wonder of the Wilderness. The name was apt, for the hotel was thirty miles from the railroad. In the early days guests undertaking the odyssey from New York took the night boat to Albany, then the train to North Creek, where the stage line started. From city to lake took about twenty-six hours, if the coach was on time. By 1886 the Adirondack Railroad had put in a through sleeping car that left New York at 6:30 in the evening and arrived at North Creek at the unseemly hour of 4:45 in the morning.

The stage left North Creek at 7 A.M., and if it was running well, it arrived at the lake at 2 P.M.

What guests found on arrival was what was often called then "the most luxurious hotel in the woods." One would suspect that arriving guests needed every comfort they could get. Every one of the three hundred rooms was equipped not merely with electric bells but with a new wonder, electric lights. The electricity of the Edison system was generated by a boiler fired with wood. On a test run some 125 lamps were burned over a period of six hours, a display of luminescence that required just a quarter of a cord of wood, which, with wood selling at twenty-five cents, cost 6¼ cents.

Rooms had running water, and a grand two-story outhouse stood in some splendor alongside the bathhouse. A steam elevator puffed up to the upper floors, and the Wonder of the Wilderness provided such divertissements as a bowling alley, shooting gallery, billiard room, and telegraph office. The minimum rate was $4 a day or $25 a week in July and August, but in 1887, in an effort to spur the early trade, there was a July-only special when one hundred rooms were offered at $18 a week. The cheapest rooms were under the main-floor porch, where, presumably, the occupants would live under the drumbeats of the strollers above; the largest suites expanded in elegance across the parlor floor.

From 1882 until 1890 Prospect House in particular and Blue Mountain Lake was the most fashionable highland resort in the northern states. It was not only grand, and because of its remoteness a marvel, it was also judged to be healthy. In support of this theory a Philadelphia physician published a sober paper on the subject in the *New York Medical Journal* of July 3, 1886. That was the clincher and the register at the Prospect House was shortly embroidered with the handsome scrawl of such names as Schuyler, McAlpin, Noyes, Cluett, Lippincott, van Rensselaer, Cleveland, Harriman, Pierrepont, Vanderbilt, Auchincloss, Colgate, Vassar, Juilliard, Stuyvesant, Fahnestock, Biddle, Drexel, Rhinelander, Astor and Roosevelt. Occasionally a visitor signed in from London, Tokyo or St. Petersburg, causing the fans to flutter even faster.

The horde pushed farther. Deep in the White Mountains

the Kearsarge House, at North Conway, found 1881 a bountiful year. Twice as many travelers arrived as had ever come before. The White Mountains had not exactly achieved social favor, but families came and so did pure-thinking people in search of mountain air, cool nights, lovely views, shaded walks and drives and moderately priced hotels. The first field meeting of the Appalachian Mountain Club would feature a paper read by a Harvard professor. John Greenleaf Whittier, the poet, who was at Intervale House for a few weeks, was expected to attend the meeting at which some plans for an ascent of Mount Washington would be discussed. Life was very outdoorsy in the White Mountains.

For more cosmopolitan pleasures one had to visit the well publicized spas, Saratoga for instance, or Newport, which by now had accumulated one hundred years of experience in entertaining summer visitors. The *Daily Tribune* came up with an unusual notion. On a mid-August day in the mid-1880s, it ran an editorial extolling the pleasures of "New York City As a Summer Resort"—about eighty years before the recent high-powered program proclaiming New York a Summer Festival. Although the papers published lists of people who have arrived at summer hotels, even longer lists could be made, the *Trib* said, of country visitors who come to New York during the same period. "They know what they're about, these country folk, and they get comfort and amusement out of the city's hotels, parks, picture galleries, and attractive shops," said the paper. "Every day, they can take their choice between a dozen ocean beaches . . . and for fifty cents enjoy one of the most delightful bay and ocean sails the world affords, hear an excellent orchestra concert, have a glorious dash in the cool surf and return to their quarters by early bedtime." And bedtime wouldn't bring any discomfort either, according to the community-minded *Tribune*. These summer visitors to New York would not "have to sleep on springless beds in six by nine rooms." Nor are they obliged "to bribe an impudent waiter to bring them a badly cooked meal."

King's Handbook of New York, a popular guide of 1893, reflecting on some of the city's leading hotels, singled out the Fifth Avenue Hotel, by then twenty-four years old, for its excellent location and fine management. It was, the book said, "unequaled in the number of and spaciousness of its corridors, halls and public rooms and the commodious character of its public rooms." Its bar was famous and, as *King's Handbook* had it, a man had to be not merely a gentleman but in "good condition" to get in. Along the route to the bar stood a cluster of maroon plush benches, called the Amen Corner, a favorite place for politicians to meet, meditate and come to momentous decisions. The Brevoort, near the Fifth Avenue Hotel, drew English visitors and is popularly accepted as the hotel Henry James had in mind when he described the arrival in New York of two Englishmen in "An International Episode." "The wide doors and windows of the restaurant stood open, beneath large awnings to a wide pavement where there were other plants in tubs and rows of spreading trees and beyond which there was a large shady square without any paling and with marble-paved walks." With the "odor of fresh flowers and the flitting of French waiters," it seemed to one of James's visitors to be like Paris. "Only more so," replied the other.

From Washington Square north to Central Park, Fifth Avenue was a residential boulevard, but there were thirteen hotels along the route, among them the Brunswick at Twenty-sixth Street, the Holland House at Thirtieth, and the Cambridge and the Waldorf facing each other across Thirty-third. The Bristol stood at Forty-second, the Windsor at Forty-sixth Street, and the Buckingham at Fiftieth. An early edition of the Plaza Hotel occupied the west side of Fifth between Fifty-eighth and Fifty-ninth with the Savoy across the street and New Netherland, sixteen stories high and the tallest in the world, on the corner of Fifty-ninth. When it opened in 1891 the Holland House was considered to be the best in the country and one of the greatest hotels in the world. It was converted into business quarters twenty-eight years later. The Buckingham, a favorite with families, gave way to Saks Fifth Avenue in 1922, and the Windsor went up in an elaborate curl of smoke in 1899.

Rather more classy then than now, Broadway had some fine hotels of its own, notably the Hoffman House between Twenty-fourth and Twenty-fifth, which *King's* called "sightly and beautiful." Down in the financial district the Astor House, built in 1836, continued as a landmark, and although it was a conservative establishment, it still drew a faithful clientele to its great rotunda for lunch.

Rockaway and Coney Island were the city's prime ocean resorts, both of which were doing well without any boost from the city's press. Reports of a Sunday at Coney noted that although the beaches, the "carousels and other catch-penny contrivances were crowded, in the sea alone, although the bathers were very numerous, there appeared to be room for more." Unless there was standing room only in the ocean, it wasn't a successful Sunday. Boats came from Elizabethport, Sing Sing and Peekskill, and one from Newark to Bay Ridge carried 683. There were a number of sermons at Coney in the morning but by afternoon everyone was in the surf and there was a great demand for "bathing clothes."

In a rather unchivalrous aside, one reporter for the *Daily Tribune* observed that the origin of so many fat women at Coney remained an unanswered problem. "There is no better way of appreciating the immensity of the ocean," he wrote, "than by noticing that no matter how many of these huge creatures roll and toss about in the water, they do not perceptibly affect the ebb and flow of the tide."

America's addiction for sea bathing proved habit-forming, and it was a long, restless wait from September until June. One of the earliest pioneers who went looking for better weather in the winter was a Rockefeller partner named Henry M. Flagler, who journeyed to Florida in 1878, bringing his bronchial wife. Although the good lady died three years later, Flagler never forgot Florida. He took his second wife there on their honeymoon in 1883. At that time all of Dade County, now the Miami area, had a total population of little more than five hundred, and Miami, Fort Lauderdale and the Palm Beaches still awaited their discoverers, unplanted with anything more civilized than scrub pine. From Savannah, travelers could take the night train, which arrived in Jacksonville sixteen hours later. There was also an inland steamboat line which was supposed to make the trip in a day, provided it didn't run aground.

After a few days in Jacksonville the Flaglers picked up a steamer and sailed thirty miles up the St. John's River to Tocoi, where they boarded the St. John's Railroad (thirteen years before, it had been horse-drawn), which for $2 took them into St. Augustine. Flagler was enchanted. Two years later he was back buying land, and by September excavation began for the famed Ponce de Leon Hotel. It opened for the January season of 1888 with 450 "sleeping apartments," suites, rooms and bridal chambers, all of them with electric lights and steam heat and some paved with Brussels carpet, hung with damask draperies, and decorated with rosewood, walnut and mahogany furniture. It cost $2.5 million and was the true progenitor of the Fontainebleau, the Doral and the elaborate hotel life of Florida that was to bloom.

While the Ponce de Leon was under construction Flagler realized that if St. Augustine would prove as popular as he expected, it would need hotels of modest cost, too. For middle-income travelers he began the Alcazar, the façade of which was a copy of the Alcazar at Seville. It had only seventy-five rooms, but it was a gem and many thought it more beautiful than the Ponce de Leon, up to then the quintessence of transplanted Spanish gingerbread. An independent hotel called the Casa Monica went up alongside the Alcazar and Flagler promptly bought it, renamed it the Cordoba and eventually joined it to the Alcazar with a bridge.

Flagler, who had come as a honeymooner and stayed to become a hotelier, now turned to transportation. He bought the railroads between Jacksonville and St. Augustine, built a bridge over the St. John's River opening uninterrupted traffic from New York to his hotels in St. Augustine. Then he began a march down the coast.

First, he stretched the rail line to Daytona. He bought a hotel at nearby Ormond Beach, enlarged it and built a golf course. When Flagler's rails arrived in Palm Beach in 1894, he had already been there a year laying out the town of West Palm Beach, the workshop for the resort, and supervising the construction of the Royal Poinciana Hotel. In 1893 Palm Beach Island, between Lake Worth and the Atlantic, had been a wilderness. In 1894 the Royal Poinciana opened with

540 bedrooms, an enormous yellow-and-white armory that for years to come would be a celebrated wateringhole for wealth, fashion and society. Eventually its size was increased until it could store twelve hundred souls a winter's night. One waiter served only four diners, one chambermaid cared for a handful of rooms, and in every hall a bellboy waited upon every whim.

Unlike the rustic resorts in the New England mountains, guests at Palm Beach were not left to amuse themselves with hikes over the countryside. For one thing, a cross-country hiker would have needed a machete. Two golf courses were laid out, motorboats were brought down to Lake Worth, bicycles and wheelchairs were provided for perambulations. Somewhat akin to the first Atlantic City strollers, the rolling carts were wicker chairs on wheels, powered by bicycling Blacks. To improve the view the design was changed so that the bicyclists drove from the rear. In effect, they were glorified rickshas which Palm Beach called Afromobiles, a name that still sticks to the few relics that roll along the edge of Lake Worth today.

Oddly enough, this, the first of the great seaside hotels of south Florida, was not on the sea; it was on the lake. Guests were conveyed on a sort of trolley car across the width of Palm Beach island, to the beach. When Flagler began his second hotel, the Palm Beach Inn, during the summer of 1895, he put it at the edge of the sea. It was to become, after a series of ruinous fires and rebuilding, the famed Breakers, which still stands today.

After the Palm Beach Inn opened Flagler launched his rails across Lake Worth, adding a bridge for pedestrians. The trains that had begun their journey in the North could now be shunted from West Palm Beach right to the door of the Poinciana, and beyond to the door of the Breakers. For the children, however, it was much more fun to stay at the Poinciana and ride to the Breakers Beach aboard the mule-pulled trolley. Hardly less important to the beach than the lifeguard was the officer in charge of decorum, who saw to it that the ladies followed the resort's stipulations that they wear black stockings connected without interruption to the bathing suit. A deviation would almost surely bring an admonition.

A huge lunch back at the hotel followed the morning swim, and through the courses an orchestra played for dancing. There was dancing again at tea time in the famed Coconut Grove. Waiters served slices of the Grove's famed coconut cake to tempt the palette while the tango seared the soul.

By 1895 Flagler had consolidated the scattered rail lines into the Florida East Coast Railway, but the sixty-six miles from Palm Beach to Miami was wilderness traversible by launch to Fort Lauderdale and mule cart beyond. By the spring of the next year Flagler panted into Miami with a small wood-burning locomotive pulling a load of building materials. The first passenger train came a week later. Miami was officially incorporated in July with a voters' register of 502 pioneers. The village had a few streets and Flagler paved them and laid out a few more. Then moving to the point of land where the Miami River empties into Biscayne Bay, he cleared fifteen acres and built an enormous hotel called the Royal Palm. Miami was little more than jungle, but for a community that was to live in the splendor of its inns, it started life in a manner to which it would soon become accustomed.

Still puffing with a full head of steam, Flagler now veered toward the Bahamas. He put down $50,000 and picked up the title to the Royal Victoria. Then he established a steamship line to carry winter refugees from Miami to Nassau. From New York, of course, passengers had long been sailing to Nassau aboard the Ward Line steamers that left the foot of Wall Street every Thursday afternoon and arrived in Nassau Sunday night or Monday morning. The Ward liners, which ran along the Jersey coast in view of the Highlands and the favored resorts of Long Branch and Cape May, carried electric bells, baths, pianos, and a load of passengers that were half invalids and half vacationists. The invalids, or valetudinarians, as they were so elaborately called in those times, picked Nassau because the weather there was said to be surer than Florida. It was widely touted as another Madeira, a tropical Mentone. A New York *Times* writer, William Drysdale, in a travel book called *In Sunny Lands*, wrote, "If there is any place weak lungs can go and be benefited by the climate I

believe that place to be Nassau. There were no invalids about
the hotel. Out of the 200 guests with perhaps two or three
exceptions, you would not have suspected there was a person
who knew he had lungs. When the ship returned everything
was all right until it got as far as Charleston, then with the
colder air coughing began again."

An English visitor wondered why more of his compatriots,
especially the sick ones, didn't flee the English winter for the
benefits of Nassau. Perhaps, he thought, they had been dissuaded by yellow fever, which had broken out on one or two
occasions, but then experts agreed the epidemic had been
blown to sea by the hurricane of 1866. "If England ever came
alive to the value of these places as health resorts," the
English visitor wrote, "the steamers would run direct instead
of requiring one to stop off at New York."

The Royal Victoria, which Flagler had bought, had been
built at the outbreak of the American Civil War when the
Bahamas, which had, since they were first settled, tried both
piracy and wrecking (and would later grow temporarily rich
running rum), now entered into the spirit of running the
federal blockade that had been laid across southern ports.
Cotton on the wharves at Charleston, Wilmington and Savannah
was worth a dollar a pound landed at Nassau only forty-eight
hours sail away. Without the South's cotton the mills of
England would close.

Full of big money and high excitement, the Bahamas
voted $130,000 to build the hotel to house their southern
friends. Some northerners had stayed there, too, but not always
with popular accord. Samuel Whiting, the U.S. consul, was
taunted as "Abe Lincoln's spy." He wrote the Secretary of
State that "Nassau is as secessionist as any southern state, aye
as any three southern states." Whiting once became so in-
censed at this hive of the Confederacy that he stalked into
the gentlemen's parlor of the Royal Vic and flailed his cane
in all directions, a lapse of temper that got him recalled.

But all that was decades past now. The hotel, which
earlier had been managed by Grover Cleveland's brother, had
emerged as one of the best inns in all the hot-weather islands.
It had a billiard room, a bar, a barbershop, broad piazzas and

rates that began at $3 a day. The Nassau band played on the lawn and the temperature idled at a pleasant seventy-five. With the New York newspapers laid out in the reading room it seemed to some like a Saratoga hotel in summer. In the courtyard captains of sailing ships suggested visits to the sea gardens and fishing excursions. Hawkers offered shell necklaces, baskets, canes, and open-lace doilies called Spanish work. Friday polo matches were a social event and between times there were frequent "matches at baseball" between the hotel and the town. On ship days the trunks were piled at the door and those who were homeward bound boarded the launch carrying flowers. "The tender," wrote one visitor, "looks like a floating rose garden as it pulls away."

The Royal Victoria, which Flagler was to embellish with another $50,000 in improvements, was by far the best building in Nassau, which had another hotel, the Curry House, and a number of boardinghouses. There were still thatched houses and one thatched church in the Bahamas, and the natives seemed only a few steps out of Africa. They were divided into Yorubas, Egbas, Ibos and Congos. Some still retained the tongue of their tribes. Some still filed their teeth and tattooed their faces. They held fire dances and beat drums and elected their own king and queen. It was only a little over thirty years since a British man-of-war had freed the last slave ship. Some of the Africans who had made that voyage were still alive. As for the Out Islands, the beach at Dun More Town was already renowned, Bimini had just given up wrecking for the more honest toils of agriculture and fishing, and the Exumas were quietly raising cattle and sheep. The natives were afraid to invade the forests of Andros Island, which were largely believed to be inhabited by Yahoos. A guidebook published in the late 1800s agreed that there might be some foundation for such a belief.

Bermuda, a two-and-a-half-day sail from New York, as it is today, was considered to be more gay and perhaps a bit more continental than Nassau. Devil's Hole was a tourist showplace even then, and for a shilling visitors could crowd around the rails to see the trained fish come to the surface in search of a handout. Hamilton was lined with wooden houses, but the masts of the ships tied up there rose high over the rooftops. The sea gardens in Nassau and Bermuda were prime attractions and there being no glass-bottom boats, ladies in long dresses, shaded by parasols, ventured out in schooners to peer into glass boxes, while natives bobbed to the surface bringing spindly black sea urchins and pieces of coral trees. The Ward liners *Cienfuegos* and *Santiago*, 2500 tons each, continued onward from Nassau to the Cuban ports of Santiago de Cuba and Cienfuegos; and while their advertisements called Nassau the Land of the Lotus and the Oleander, one illustration, which might well have dissuaded many a matron from making the voyage, depicted an enormous crocodile chasing a native boy up a palm tree.

Most passengers who ventured further into the Caribbean sailed on private yachts which called at Anguilla, St. Martin, St. Bartholomew, St. Eustatius, St. Kitts, Antigua, St. Lucia, Montserrat, Nevis and Grenada, even poking into Trinidad and Surinam. Passengers escaped the heat by rocking in hammocks under a canvas awning, by tying themselves to safety lines and jumping in the sea, and by soaking themselves with hoses. In port, most of them stayed aboard their ships, for the only islands with suitable inns were Barbados and Trinidad. Barbados, where Washington had visited 150 years before, had two hotels, one rather oddly called the Ice House and the other the Marine Hotel, a favorite winter retreat for Americans which was managed by a refugee from Maine. The Ice House was the prime resort of the islands, with the bill of fare, usually featuring flying fish, prominently displayed on a black board hung at the entrance. Tissue-paper telegrams telling of news that had happened around the world fluttered in the public rooms. It had only twenty rooms, but they were about half the price of the best hotels in Long Branch and Saratoga, and then, of course, there was the added embellishment of the Swizzle, said to have been an Ice House invention. Gin, lemon and soda were the ingredients, mixed of course with a swizzle stick, an eighteen-inch-long peeled bark wand, thick as a lead pencil and flowing with colored ribbons.

Across the continental United States, meanwhile, there

gradually grew a network of hotels and resorts that brimmed
with the latest improvements, the refinements, and the ultimate
in fussy decoration. The United States Hotel in Boston bragged
of its recent renovation, told its prospective patrons that it
was free alike from extravagant show or still more extravagant
charges. Its location was so ideal that patrons could save all
carriage fares. Horsecars pass three sides of the hotel connecting with railway stations and steamboats. Kansas City had
the Bonaventure, a new family hotel with 130 rooms to let
at $3 to $4 a day. Every room was fitted with incandescent
light, the suites had private baths, and from the outside,
minarets and gabled arches and bay windows lent the façade
a brilliant, if busy, aspect.

But the marvel of the Midwest was surely the Palmer
House, which, after two years of labor and $2 million in con-
struction costs, opened in 1873. It had seven hundred rooms,
and Potter Palmer, a Chicago millionaire, had spent half a
million in furnishing them. Naturally, the architecture and
design were European, Mr. Palmer and his architect having
conducted a continental tour, putting up at the best hotels and
skimming the best ideas. Although the management insisted
the Palmer House was original in every detail and in all its
eight stories, it did admit that the exterior was remindful of
the Hôtel de Ville in Paris. The State Street entrance was a
Grecian portico, its canopy upheld by huge statues representing Science, Agriculture and Commerce. Potter's press agent
alleged they were the costliest works of art ever used to
ornament a building in the United States. Moreover, there
was more brick in the Palmer House than in any two other
hotels in the country, more iron than in all the nation's hotels.
It was so fireproof that Palmer invited other hotel owners to
try to build a fire in any apartment in his hotel if they would
permit him equal privileges in theirs. The offer was never taken
up, but Chicago was extremely conscious of combustion. Only
two years had passed since Mrs. O'Leary's cow kicked over
the lantern.

Not to be outdone by the splendors of Newport, Saratoga,
Bar Harbor and White Sulphur, the Midwest had a gilded
playground of its own, the famed Hotel Lafayette, situated,

not altogether handily, fifteen miles west of Minneapolis. A railroad baron, this time James J. Hill, personally picked the location for this, the first hotel of the St. Paul, Minneapolis and Manitoba, a line later to be better known as the Great Northern Railway. The Lafayette opened in 1882 with so many bookings—a thousand, it was said—that it was enlarged the next year. It stretched for 745 feet along the lakefront and was 95 feet wide. Its floor area—hotels were being built out, not up—covered five acres and every one of its three hundred accommodations was a front room looking at Lake Minnetonka.

From the beginning, the Lafayette was a smashing success, attracting not merely surrounding midwesterners who thought it too far to trudge to Saratoga, but southerners who came up by the steam packets of the Diamond Joe Line that sailed from St. Louis to St. Paul. Scores of steamboats huffed about the lake and trains ran hourly from St. Paul and Minnesota. The overflow went into the Lake Park and the smaller St. Louis, but the Lafayette was the center of social life on the lake, and natives were not invited. Professor George Seibert's orchestra played for hops three times a week, and betweentimes, for those who found themselves deficient, Miss Kent and Miss Barry held afternoon dancing classes. Open-air military band concerts blared in front of the 1200-foot-long porches, Negro cakewalkers and minstrels picked from among the employees filled in at show time. Fireworks exploded every Wednesday and the daylight hours were filled with regattas and tennis. A German correspondent, on hand for General Grant's second visit in 1883, recorded an early September scene:

> Out on the beautiful lawn overlooking Lake Minnetonka, I see from my window Lord Carrington in earnest conversation with Lord Onslow. At a little distance from them, Hon. James Bryce, the historian, has buttonholed Dr. Burchaardt from Manchester. They are engaged in constant conversation, hardly noting the glory of the sunset. . . . President Arthur, chief executive of the United States, is expected tomorrow.

Like many another giant summer hotel, the Lafayette, alas, perished in flames long before its vogue had waned. It burned to the ground in 1897, and the total value of the property rescued from the blaze was estimated at $50.

In the era of the Lafayette, railroad travel was inexpensive and frequently sumptuous. A half section in a sleeping car between New York and Chicago cost $5, and from Chicago to St. Paul, another $2. The Northern Pacific, which hooked dining cars on all through trains, served splendid dinners at a flat rate of 75 cents. Beginning at St. Paul, Minneapolis and Duluth, the Northern Pacific edged into Yellowstone National Park in 1881. It carried parlor cars between Fargo and Bismarck, where only ten years before surveyors had found all their stakes pulled, bundled and tied to the existing railway line. With its routes spreading through such fertile territory, the NP had its eye out for all sorts of customers. "It responds," one of its ads went, "to low rates for land seekers, settlers and tourists."

The land seekers were fanning out in all directions, but the tourists, what there was of them, rode the rails as far as Cinnabar, Montana, where stages rumbled onward to the Mammoth Hot Springs Hotel, the center for visitors to the national park. Ever since the Indian resistance faded, Denver had begun to find itself something of a resort as well as a city. As it turned out, the vacation place became Colorado Springs, seventy-five miles south, sitting on a six-thousand-foot-high plateau in the shade of Pikes Peak. Colorado Springs had clear skies, no fog, dry air, bright sun, little rain. In fifteen years it had become the sanatarium of the Rockies, an attractive tree-lined city with a devoted and rather chic coterie of two worlds. By 1883 it had a fine hotel called the Antlers, which sheltered what *Harper's Weekly* called in 1886 "cultured Americans and Englishmen whose chief motive in a change of residence was to regain health. . . ." Said the magazine, "In artistic and literary refinement, Colorado Springs without doubt leads any city west of St. Louis. . . ."

As the rail lines reached the Pacific and then began to fan out in all directions, some new and rather entrancing notions were finding their way back east. Southern California, that strip from Santa Barbara to San Diego, had long been known, albeit rather vaguely, as an all-season retreat for invalids. Although they were only accessible by sea, Los Angeles, San Bernardino, Anaheim, San Diego and Santa Barbara had been visited, and some had even sprouted fine hotels. When the Southern Pacific extended a leg six hundred miles south and east of San Francisco in the late 1870s it brought all of Southern California within a train ride of the rest of the country. Said the New York *Herald* one February day in 1877, "The expense and time consumed in the trip are but little more than a trip to Florida, and of course, very much less than the trip to Cuba or the Mediterranean; while it embraces many advantages which none of them afford." The train trip from New York to Southern California took seven and a half days, but there were visionaries, the *Trib* among them, who saw a great future under the orange trees. Said the paper, "It is the prediction of many who have visited Southern California . . . that it will become the great natural health resort of this continent."

On its way to fulfilling that prediction, Southern California would surely ruffle many a tourist spoiled by the sophisticated comforts of the East. By 1886 New York papers were advising visitors to spend at least a week in Los Angeles since it was the hub of so many interesting colonies. But the traveler was warned that there was no good hotel in Los Angeles, unless one considered the inn at the depot, where the food was acceptable but the noise of the trains insufferable. Private boardinghouses took advantage of the shortage and offered quarters for $10 to $15 a week, but in the side sheds rooms could be rented for $3 to $6 a week. Cheap American-style restaurants charged a quarter for a meal, but one had to expect to pay double that for a French *table d'hôte* dinner. For the brave and the pioneers there were the lowbrow Mexican places around the depot. A traveler who picked carefully could get by in Southern California for $2 a day, but in order to see the town one would have to expect to pay another $2 an hour for a coach and driver. A drive-yourself single rig could be rented for $4 to $5 a day.

Thomas Cook and Son, the British travel agency which kept in touch with its clients by publishing an ambitious newspaper known as *Cook's Excursionist and Tourist Advertiser,* found itself enchanted with the possibilities of California. In its issue of December 1888, just in advance of the winter season, the *Excursionist* gushed rapturous paragraphs about this new and toasty Valhalla. "It is doubtful if any instance could be adduced that would in any considerable degree compare with the stupendous growth of Southern California during the past five years," the *Excursionist* opined, albeit in a rather crusty piece of prose. The thousands of miles of new rails, the new cities and the fruitful soil all made California a new wonderland, but, as the *Excursionist* pointed out, it was the "superb climate that gives it the position of the great Sanitarium of this country, and the finest winter resort for the invalid or pleasure seeker."

To allay the reputation which had made its way east over the Rockies, the *Excursionist* was, some paragraphs later, at pains to note that considerable improvement in accommodations had taken place during the previous five years. The Nadeau, the Westminster and the new United States had all appeared in Los Angeles, and the Arcadia at Santa Monica got Cook's blessing, too.

But the most imposing of all was the del Coronado at San Diego which had opened the prior February with 750 rooms spread over seven and a half acres. In keeping with hotel architecture of the times, which seemed to imply a ratio between elegance and length of frontage, the del Coronado stretched for 1300 feet. The rest of it was a parallelogram that enclosed a grand court 250 feet by 150 feet, all of it watched over by an observatory that rose 150 feet. It took thirty billiard tables, four bowling alleys, and an unrecorded number of shooting galleries to provide indoor amusement for its guests, not to mention lectures, the minstrel shows, the stereopticon showings of the splendors of Yosemite, the dancing classes for children and the darkroom for those who were taking snapshots with the Kodak. Outdoors one rode to the hounds, fished in the sea, went yachting and rowing, played polo and hunted for rabbits, quail, snipe and ducks. Nowhere in Europe or America were there so many things to do in a place so large.

Los Angeles, the Cook's reporter noted, was too well known to need a description in its pamphlet of 1888. By that time it already had a population of sixty thousand. The Cook's chronicler dismissed it by saying that it was merely the "Paradise of America," with the climate "as near perfection as can be." The city, it said, was brilliantly illuminated with electricity; electric motor lines and horsecars had been introduced; and flowers and fruits abounded everywhere. "Facilities for traveling in Southern California," said Cook's, "have been greatly advanced within the last few months." One of its homegrown competitors, Raymond and Whitcomb, published a whole treatise called *A Winter in California* and described this mode of travel in great detail. "The vestibuled train . . . ," it said, "is united under one continuous roof, so that in place of detached cars, with exposed platforms, there is in reality an elongated suite of elegantly furnished apartments. . . . There is also a well selected library of standard books of travel and fiction, free for the use of all passengers. . . . Another part of the car is devoted to the barber's shop which is under the charge of an experienced and polite attendant. . . . Adjacent to the barber shop is another novel feature—a bathroom. . . ."

In this commodious style the first-class traveler could ride the rails southbound to Mexico or north to San Francisco and Monterey. Northern California was already well known. The Palace Hotel in San Francisco, which had opened in 1875, a mere quarter of a century after the gold prospectors came bursting into the undistinguished town, was a $5 million spectacular. It had 755 rooms, covered a city block, separate dining rooms for breakfast, for children, for ladies. It had reading rooms and barbershops, public baths and billiard rooms, slots on every floor to drop letters, pneumatic tubes to drop packages. It claimed to have started the idea of positioning a clerk on every floor ready to communicate messages to the desk through a speaking tube. The idea was later adopted by the Palmer House and the Grand Pacific in Chicago and, reversing the direction of most trends, spread eastward to New York.

But the most engaging sight of all San Francisco was the Grand Central Court of the Palace, a version of an old-time courtyard inn covered with a crystal roof, bursting with a tropical garden, bubbling with fountains, paved with marble, and resounding, every afternoon and evening, to the sweet timpani of a band. If this extravaganza was a mere splendor by day, it was a phantasmagoria by night, when 155 stands of gas fed 503 multicolored lights.

Despite the Palace Hotel, which had just opened when he arrived, Anthony Trollope, homebound to England from a trip to Australia, was utterly bored with San Francisco. "There is nothing to see," he said, "worth seeing. . . . There is a new park which you may drive for six or seven miles . . . there is an inferior menage of wild beasts, and a place called the Cliff House to which strangers are taken to hear the seals bark . . . and the ordinary traveler has no peace left in him either in public or private by touters who wish to persuade him to take this or the other railway route into the eastern states." A decade later, a travel book by Ben Truman published in New York found visitors still looking at the Palace Hotel, then the "cable roads," the "Chinese quarters," Golden Gate Park, Russian and Telegraph Hills, the Presidio, Cliff House, the Oakland Ferry Building, and the Safe Deposit and San Francisco Stock Board Buildings. "The Palace Hotel amazes all visitors," Truman wrote. "A bath and a closet adjoin every room . . . the rates are room with board, $3 per day; room with board, $4 per day; room without board, $1 per day and upwards. . . ."

Truman was not nearly so lyrical as the New York *Daily Tribune*, a continent away, which told its readers that anyone boarding a cable car in the fabled city by the bay had "a noble ride in front of him." Its notes on the ride were a small travelogue on the city:

> The view is unobstructed; the motion is even and regular; the frequent stoppages cause no annoying jar or jolt; the track is cleared with great promptness by all vehicles; one sees the busy life flowing up and down on either side, but has no share in it and is not jostled by it; it surpasses

in comfort and exhilaration a ride downtown in the coach seat of a Broadway stage. The car sweeps up the broad avenue, past the huge, many-storied, bulbous-windowed, factory-like Palace Hotel, the ugliest as well as the largest in the world; past the rows of stores with their big show-windows filled with costly goods; past the Baldwin Hotel, one of the most imposing buildings in the city, past the New City Hall, a dust-covered, dull-red brick structure, of peculiar architecture, dwarfed and shorn of its little beauty of stature and proportion by the ugly, shambling structures built around it, on the historic sand-lots, which a greedy and unscrupulous city government sold out at auction to the highest bidder; past this monument of municipal extravagance and official meanness to the sharp curve which marks entrance on Haight Street. The cars round the turn with scarcely a jolt. Now the rider's face is set squarely toward the west and his course is a bee-line for the ocean beach, near the famous Cliff House. Up a gentle hill the cars climb without apparent effort. Everything here has a painful air of newness, and well it may for the street has been created within a year.

In its asides to its customers, *Cook's Excursionist* could scarcely have too much to say about the glories of the Palace Hotel, where it boarded its trippers and maintained an office. But it also wondered aloud whether there was anywhere else in the world where, among a population of three hundred thousand, there existed so many opportunities for what it called "pleasant thought and study." One of the most interesting of the city's phenomena was the Chinese quarter. "They are huddled together," the Cook's man wrote, "in a rectangular space seven squares in length and by three or four squares in breadth, right close to the business centre and the palaces of the railway millionaires. In fact, one who (like the writer) has visited China, will feel in 'Chinatown' that he has been transferred suddenly from America to China, the only drawback to the illusion being that the character of the buildings are different to those met with in China proper."

For the visitor, not the least of San Francisco's wonders were its restaurants—all sizes, all prices, all degrees of sanitation. Wrote a New York reporter in 1880, "St. Louis, Cincinnati, Baltimore, Albany, all of these cities have a few fine restaurants where the wayfaring man, by squandering from $2 to $10 may get a good dinner. But in none of them will be found eating houses scattered through the city, and graded to suit the tastes and purses of a large number of people. Even in New York where the cuisine of every nation—save perhaps that of the South Sea Islander—is represented, one fails to find the variety and cheapness which are the two main characteristics of San Francisco."

By cheap the writer meant five- and six-course French dinners with native wine for 25 cents. The same in New York would cost at least 40 cents and one could count on an acid *vin ordinaire*, at an extra charge of a quarter. Said a French-woman to the reporter, "I was desolated in your Boston; I wearied myself to death in New York where the streets on Sunday were like the tomb and the theaters were all *fermés*; but when I reach San Francisco, ah, it was heaven, it was a leetle Paris!"

From San Francisco hardy excursionists might undertake the Yosemite Valley; those in search of care and comfort slipped down to Monterey. Trollope, who made the trip into Yosemite, might well have wished he had chosen Monterey instead. First, there was the rail ride to Merced, but after a 140-mile ride it was time for bed. "Hotel comfortable, although they refuse to clean boots. The four-coach stage left at six the next morning and rumbled over the roads at five miles an hour. After twelve hours passengers were more than ready for a halt, even at the ranch, where lodging was provided," Trollope reported. In September, when Trollope made the trip, the riverbeds were dry and the dust was unbearable, but in May and June he imagined the aspect might be more felicitous. "In those months, however, the place is full of travelers; the ladies sleeping in dormitories and the men either under or on the dining room tables," he wrote.

Monterey, which had been a city under the Mexicans, was a port of comfort. Many old Californians had looked upon it

as their most healthful resort, though with its rather frequent
fogs and chill winds off the Pacific one scarcely knows why.
After the Hotel del Monte was opened in June of 1880 it
became a celebrated station for Americans and Europeans as
well. It had 240 rooms all with hot and cold water and the
use of the bath on the floors was provided free. Set in a
beautiful park, its gardens, avenues and walks were maintained by a crew of fifty. Sandboxes were set aside for the
children, who couldn't wait for the daily visit to the beach.
The bar, bowling alley and smoking room, all private purlieus
of the men, were in a separate building known as the Club
House. This rather idyllic life by the sea in Monterey's del
Monte cost $2.50 a day.

If California was for adventurers, what lay beyond was
for explorers. Those who ventured across the Pacific to Hawaii
could find rooms at the Hawaiian Hotel, a plain wooden
building sitting on a handsome banana-bedecked lawn three
miles from Waikiki. When it was built in 1872, it cost $175,000,
and if that was a modest figure alongside some of the palaces
that were being built on the mainland, the rooms were large
and airy and each had its own veranda screened with mosquito
netting and gauze. All of the servants were Chinese and they
fetched a breakfast that could include poi served two ways:
one like mush, with sugar and cream, the other pressed hard
and eaten with butter and salt—or so reported a visitor in 1898.

The Hawaiian Hotel maintained a Waikiki Villa on the
beach, where guests could either rent suits or store their own.
There were dressing rooms at the edge of the sand, and a
toboggan slide built high above the shore propelled guests
down the chute and far out into the combers. Visitors were
taken off to see the sugar and pineapple plantations at Pearl
River and at Waianae, to the Pali, to a dead volcano known
as the Punch Bowl, and to the other islands. At Volcano
House, an inn built at the edge of a crater on the island of
Hawaii, the proprietors had built a bathhouse near a fissure
in rocks and lava. Steam from the volcanic crack was fed into
the bathhouse, allowing for "vaporous baths." In the evenings
the guests sat around the fireplace and listened to the tales of

earthquakes and lava flows while, as one 1892 guest later wrote, "the glare from the volcano lit up everything outside the house."

Although mosquitoes were a great problem in Honolulu, it didn't seem to impinge on a lively social life, much of which was centered about the Hawaiian Hotel. On Sunday afternoons a stream of carriages bounced to the athletic grounds, where gentlemen resident in Honolulu formed teams to play baseball. But it was Waikiki, "the seaside resort of Honolulu," that excited the visitors. One Helen Mather, who spent a summer in Honolulu and wrote a book about her experiences, declared with some ecstasy, and perhaps a little hyperbole, "It's Long Branch, it's Newport, it's Deauville."

Still, the glories of Hawaii remained largely undiscovered and unsavored. Thomas Cook, which began its round-the-world tours with a dauntless party of nine in 1872, and which was now highly organized and proficient in the ways of global travel, gave it little notice. An elaborate booklet which the firm published for the Chicago Fair makes no mention of the islands at all. It recommended trips to Australasia, where "travel to the colonies is due largely by steamer," and talked with knowledge of Melbourne, Sydney, Tasmania and New Zealand. A traveler who put up at the Grand Hotel in Yokohama in 1898 reported spending most of his time there disentangling himself from the merchants, peddlers and tailors —shades of Hong Kong in the 1960s!—who came knocking on his door. Travelers with inside connections tried to get "imperial passports" which permitted the bearer to see parts of Japan which were otherwise off limits to foreigners.

Trips around the world were now a commonplace occurrence, it conceded, and while such a circumnavigation could be accomplished in ninety days, six months was better, eight if you included Australia, and nine for those who threw in New Zealand. Nagasaki, Hong Kong, Canton, Singapore, Colombo, Darjeeling, Agra, Delhi and Bombay were already labels to be prized and pasted. The Northern Pacific's ships sailed to Hong Kong from Portland for 28 pounds or less than $150. A traveler who sailed on the S.S. *China* before the century turned found the dining salon cooled by two Chinese who pulled the cord of long silken punkahs. The officers had set up a canvas tank, sixteen feet long, which they filled with fresh seawater daily, First class passengers were invited to swim.

But these embellishments would only be considered makeshift and make-do in the Atlantic, where an elegant ocean crossing was rapidly becoming the fashion. In the face of ever increasing competition and more and more luxury, the Cunard Line was required to forsake its early slogan, which offered "a plain cabin—nothing for show." Now, like the palace hotels that had risen in the big cities from coast to coast, *everything* was for show: first electric bells in the cabins, then electric lights in the saloons and finally—magical as it was— electric lights to be turned on and off in one's own stateroom. Soon the first liner would appear with—as one London hotel put it when it installed the latest fad—"constantly ascending and descending electric chambers."

Although the *Great Eastern*, which was launched in 1858 was to hold the heavyweight championship of the Atlantic for forty years, the *City of New York* and the *City of Paris*, launched in 1888 and 1889 by the Inman Line, were the first express liners to be fitted with twin screws. The Cunard Line produced the 12,950-ton *Campania* and the *Lucania* in 1896 and sent them scurrying across the Atlantic at twenty-one knots. A year later the Germans were in the race with the Norddeutscher-Lloyd's *Kaiser Wilhelm der Grosse*, fastest liner in the world. In a year's time it stole a quarter of all the Atlantic business. Cunard and the White Star Line met the challenge, White Star countering first with the 17,300-ton *Oceanic*.

Despite her size, *Oceanic* could not match the speed of the *Kaiser Wilhelm*. The White Star Line declared that it had sacrificed a few hours of speed for comfortable passages. Hadn't the public already noted how the Germans had driven their ships through the most miserable weather? Hadn't the engines been strained, even laid up at port while repairs were made and schedules remade? The *Oceanic* would have none of this. Six days between the Old World and the New were fast enough, and the duel ended, for a while.

The glory of the United States in these transatlantic races was, for the twenty-year period from 1873 to 1893, upheld, if that's the word, by a rather undistinguished ship service that plowed between Philadelphia and Liverpool. In 1893, however, the American Line took over the *City of New York* and the *City of Paris* from Britain's Inman Line. The U.S. passenger fleet was further enhanced by the *St. Paul*, which, while neither the fastest nor the largest ship afloat, was the first to have a ship-to-shore wireless and the first, glory be, to publish an ocean newspaper.

In a notice printed late in June of 1881 the New York *Daily Tribune* reported that reservations aboard transatlantic steamers were at a premium. The increase was a good 20 percent over 1880 and even the captain's and officers' rooms were being requisitioned. The rush to Europe was partly attributable to the national prosperity, the *Trib* said, but also to the agreeable fares. A trip to Europe, said the paper, cost little more than spending the same time at Saratoga, Newport or Long Branch. A first-class cabin on the Cunard's *Campania* and *Lucania* cost $90 to $150, second class cabins $40 to $50. On the *Umbria* and *Etruria* first cabin rates started at $75, second class $35. Voyagers could take their servants along at a flat $50 fee. German ships were about $10 less in all classes than the Cunarders.

A crossing in steerage class cost $15, and was usually worth just about that. But as more and more velvet appointments were fitted into first class, some minor refinements found their way down to steerage, too. Even as early as the 1880s some young adventurers, eager for a look at Europe, took their bikes along, content that they could endure any discomfort for six or seven days. In 1895 Cunard made an outright move to encourage steerage travel by urging young people to organize European bicycle parties.

The growing tourist trade to Europe, which began almost immediately after Appomattox (the New York *Times* estimated that there were 100,000 Americans in Europe either as residents or as travelers in the summer of 1865), produced all sorts of advice and comment from editorial writers to guidebook authors. The New York *Times* in 1866 noted that Paris was drawing the greatest number of Americans, but that the German towns were becoming great favorites, too. Americans were attracted to Germany by "the cheapness of living, the cultivated society, the comparative quietness of the course of life." Many Americans, the *Times* observed, were giving their children the advantage of German "educational gymnasia and colleges which abound there." The number of Americans, both resident and transient, coming to Rome now formed, the *Times* said, an "important feature of the foreign population."

Lippincott's Magazine, in a review of *A Satchel Guide of the Vacation Tourist in Europe* in 1872, warned that the traveler who " 'bolts' Europe like a meal in a Bowery stand-up restaurant" was now beginning to create a literature of his own. The *Lippincott's* reviewer disdained the "American tourist of the present time"—perhaps the "clerk who has exactly sixty days and forty-eight minutes of absence . . . from Sandy Hook," the landmark at New York harbor, before he must return home. The "little book carries its chin very much in the air," chided the *Lippincott's* reviewer, and is "own kin to the young man from America . . . whom we send over to the Continent in singlets and brass heels or in rattan and short jacket, to conquer the Old World he is just too big to cry for."

Despite the acid review it received from *Lippincott's*, the *Satchel Guide* had many a nugget of knowledge for the "vacation tourist," who, as it said in the preface, "can spend but three or four months abroad." Among the hints: carry ten English sovereigns to cover shipboard expenses and the first few days in England; for shipboard bring a "comfortable winter suit that has seen its best days, an old thick overcoat, with a traveling shawl or 'rug', and a soft felt hat"—all this to be left in an old carpetbag or other shabby piece of luggage at Liverpool and to be picked up on the way home; once in England buy a good English valise for the trip itself; bring a passport as a useful means for identifying oneself at banking houses or art galleries even though you will not be asked to show it in Great Britain, Germany or Switzerland, and perhaps not in France or Belgium.

Learn about architecture and some smattering of languages

before you go; "do not indulge in seasickness, or if you must be seasick do not weakly succumb to it, but fight manfully against it"; when going through customs be advised that a single copy of an American reprint of an English copyrighted book is liable to be confiscated; candles and soap are extra charges in European hotels; expect the number of your room to be sewn on your clothes in black thread before they are sent to the laundry; on shipboard give the table steward five shillings, give the room steward two thirds of what you gave to the table steward, give the bootblack a shilling. Despite this liberal seeding of the help, a four months' tour could be done for $400 including a return ticket aboard the moderately priced National Line.

Americans stayed at the Hôtel d'Italie and Bauer, also known as the Bauer Grunwald, a 200-room establishment on the Grand Canal complete with Ladies' Dining Saloon and Wine and Beer Cellars. In Monte Carlo they were at the Grand Hotel Victoria, 350 rooms and a full south aspect, an attribute highly prized in the time. At Menton they were at the Grand Hôtel de Venise surrounded by a garden and handy to the English Church. The Baur au Lac in Zurich, one roamer wrote, "exceeds in romantic situation, princely elegance, and home-like comfort any other that I've seen in Europe not excepting the Grand Hôtel at Paris . . . an excellent band discourses harmoniously every day to a delighted audience of guests when seated to dinner at three long rows of tables. . . ." It became a fashion among the English and the Americans who spent the summer at St. Moritz to come to the Beau Rivage at Ouchy near Lausanne for a few weeks before returning home. When the beautiful daughter of a visitor from abroad snagged the Duke de Choiseul-Praslin, the hotel became a resounding success. It was moved to replace its candlelit salons with gaslight and in the 1890s with electricity. The hotel heard whisperings from gadget-minded Americans that private bathrooms were being installed *alongside* bedrooms, and was moved to try the idea. Americans wandered through Europe wondering when Paris ever went to bed, where all the wine came from, and why Europeans seemed so disinterested in the affairs of the United States.

To one chronicler, Lucerne was already a tourist trap, Interlaken already a fashionable resort, Baden Baden plagued by frightful gambling, Heidelberg filled with belligerent students, Vienna lulled by a dreamy languor "as if the people imbibed an infusion of poppyheads for tea." The magnificence of the railroad depots awed them, their railway carriages with half the passengers riding backward annoyed them. But nobody missed the ritual at the Rigi, a Swiss mountain which one *had* to see at sunset and again at sunrise. The hotel sounded its Alpine horn at four in the morning and the dutiful tourists, fretful lest someone should query them about it back home, trudged up the mountainside to see the brilliant view often somewhat obscured by clouds, rain, fog or bleary vision. One tourist, of the hordes who came, penned his sentiments in verse:

> Seven weary leagues up hill we sped
> The setting sun to see
> Sullen and grim he went to bed
> Sullen and grim went we
>
> Nine sleepless hours of night we passed
> The rising sun to see
> Sullen and grim he rose again
> Sullen and grim rose we

With the Rigi carefully etched in one's book of remembrances, there was the Eiffel Tower to see, a new marvel built for the Paris Exposition of 1889. Stuck among its iron girders from the beginning was a Russian Tea Room and an Alsace-and-Lorraine café, booths and sideshows, an office of *Le Figaro*, where tourists who jotted down their names and addresses would find them reprinted in the evening edition. Rome was something of a discovery and those who went found it filled with shops, their windows stuffed with colored silk scarves, delicate gold work, polished oxhorn, carved coral, cameos and mosaics, small bronzes and sculptures. Men and boys with baskets of tortoiseshell waited on every corner.

The tourist world beyond Rome seemed to be the private

preserve of Thomas Cook and Son. Their tours to Palestine had been operating almost since the close of the American Civil War. By 1893, when Cook's unveiled a lavish exhibition at the Chicago World's Fair, they had conducted fourteen thousand tourists through the maze of the Holy Land and their list of clients included Prince Albert Victor and Prince George of Wales, General Wallace, the American ambassador to Turkey, the princes of Denmark, Sweden and Norway, the grand duke of Austria and the duke of Genoa, each of whom traveled with his own suite. Jaffa had a poor reputation among travelers, and everyone scurried onward to Jerusalem, thirty-six miles away, stopping sometimes at Ramleh, where one had a choice of Nohnenberger's Inn or the Latin monastery. Dimitri's was said to be the "only hotel worthy of the name" in Damascus, which was reached from Beirut by a hard surface road. Three-horse carriages paid a toll of $24 for the round trip.

Egypt, particularly, seemed to be in the undisputed hands of the men from Cook's, who had engineered an exclusive contract with the khedive to operate steamers on the Nile. With a flotilla of five ships, Cook's offered prices for the voyage to "Assouan," as it was spelled then, for anywhere from £23 on a mail boat to £65 on a "special steamer." But it also had paddle steamers for special parties, dahabeahs, which were said to be "very popular with invalids who enjoy floating about the Nile, and with wealthy people who wish to travel in the privacy of their own circle or party." The steam dahabeah *Nitocris* was the ne plus ultra of Nile excursions. It was "adapted for invalids or shooting expeditions" and carried a dragoman equipped with the necessary authority to receive fresh vegetables, poultry from the Cook's farms at Luxor. Cook's operated both hotels at Luxor, considered a refuge for those "unable to battle with the severity of the English winter" and who sought relief from such debilities as "phthisis, Bright's disease, kidney and bladder trouble, rheumatism, asthma, as well as convalescents from typhoid."

One writer of travel letters to the Medford *Mercury* noted in 1876: "The Egyptian fellah or peasant considers the Khedive the greatest man and Cook the next. On the Nile Cook certainly is monarch." In 1891 the cost of a twenty-day excursion up the Nile, including donkeys, dragomen, side saddles for the ladies, a look at the ruins of Memphis and onward to Thebes, to Karnak and Luxor, and the quarries of Assouan, was $250. It might sound expensive considering the cost of living in the land, one travel book writer noted in 1891, but considering the saving in haggling again and again for donkeys and dragoman, the price, he decided, was cheap after all. Mary Thorn Carpenter, who wrote in Cairo and Jerusalem for the turn-of-the-century tourist, said simply, "Cook's owns Egypt. . . " It was plain, she said, "how implicitly the Khedive has confided to him the river Nile, where the flag of Thos. Cook streams from a perfect flotilla of side wheelers and even flirts from the mail boats despatched to the second cataract and beyond." Even the fellahin were wearing "blue Salvation Army jerseys with Cook and Son across the breast in red letters." The Cook conquest seemed so to awe the travel writers that they almost forgot to describe the glories of Egypt itself. Baedeker, of course, permitted itself no such flighty lapses. It painstakingly listed seventy sights to see in Cairo, then divided them into six hardworking days of eight hours each, running the dedicated through a maze of mosques and pyramids and bazaars to the tomb of the Khedives and the Virgin's Tree, an old sycamore where the Virgin and Child were supposed to have rested during the flight from Egypt.

While Cook's paddlewheelers threshed the Nile and Cook's tourists bobbed across the desert on donkeys and camels, back home in the machine shops a band of daring mechanics were working on a gadget that would do away with the horse. By 1896 there were a grand total of sixteen cars in all the United States. Like anything new, they were enough to provoke controversy in some quarters. Said an angry writer for a farm paper, "These noisy, smoke stinking wagons are designed to frighten to death anything they can't flatten out." In the years to come they both frightened and flattened, but they were also destined to revolutionize the ways and means of the man on the move.

Stagecoach was the mode of travel in the early days. Travelers boarded the Good Intent Line, the Oyster Line (which also carried oysters), and the Shake Gut Line, which, people said, lived up to its name. Boston to Savannah was a three weeks' run. Boston to New York took four days and cost $10. Dickens complained that coaches had never been cleaned.

TROY, BALLSTON
AND
SARATOGA,

DAILY LINE OF
COACHES.

This line will commence running on the first day of July, leaving each place at half past 8 A. M. every day. Passengers wishing to travel from Saratoga to Lebanon Springs, will find this line not only the most expeditious but cheapest.

Passengers for Pittsfield, Northampton and Hartford by taking this line will dine at Troy, lodge at Pittsfield, and arrive at Hartford early the next day. The road is now put in the best order, and all that is now wanting is that liberality which the establishment merits.

☞ Seats taken at G. W. Wilcox's, York House, Saratoga, and at all the Principal Houses in Troy.

L. V. & J. B. REED, Proprietors.
J. S. KEELER, Agent, Troy.
S. DEXTER, Agent, Saratoga.

TROY, JUNE 25, 1834.
N. B. On the arrival of the ERIE or CHAMPLAIN, Parties can be accommodated with coaches to Saratoga or Ballston the same evening.

Printed by Kemble & Hooper—Troy Budget Office.

Steam Ferry.
24th April, 1836.

THE STEAM-BOAT
GEN. GREENE,
CAPTAIN DAN LYON,

WILL run until further notice in the following order, viz:

Leave Burlington at half past 8 o'clock in the morning, Sundays excepted, touching at Port Kent, and arrive at Plattsburgh at 12 o'clock.

Leave Plattsburgh at 2 o'clock P. M., and PORT KENT at 4 o'clock, and arrive at Burlington at half past 5 the same evening.

The following are the established rates of Ferriage
TO AND FROM PORT KENT.

Every four wheel pleasure Carriage on springs, drawn by two Horses, including driver,	$2 00
Every two wheel pleasure Carriage on springs, drawn by one Horse, including driver,	1 50
Every Wagon or Sleigh drawn by two Horses, including driver,	1 50
Every Wagon, Cart or Sleigh drawn by one Horse, including driver,	1 25
Every Cart drawn by two Oxen, including driver	1 50
Every additional person, Horse or Ox,	50
Every foot passenger, (children under 12 years of age, half price.)	50
Cattle in droves, each	25
Sheep and Hogs in droves, each	6
Parties of pleasure going and returning the same day, not less than 12 persons, each	25

A reasonable sum will be added to the above prices to and from Plattsburgh.

The above rates will be charged, until the first day of November, after which time the company reserve to themselves the right of charging those rates of ferriage which are established and allowed by law.

Poster for steam ferry puffing between Plattsburgh and Burlington displayed price list for sheep and hogs in droves (6 cents), two-horse sleighs ($1.25), and "parties of pleasure" at 25 cents each. Tow boats, below, pulled by mules, were used on the Virginia tidewater canals and along the Erie Canal in New York.

America got into the transatlantic steamship service in 1816. Fifteen days was a record run between New York and Liverpool. The ship carried its own milking cows. The Continental Hotel, below, with its long veranda was a Newport favorite.

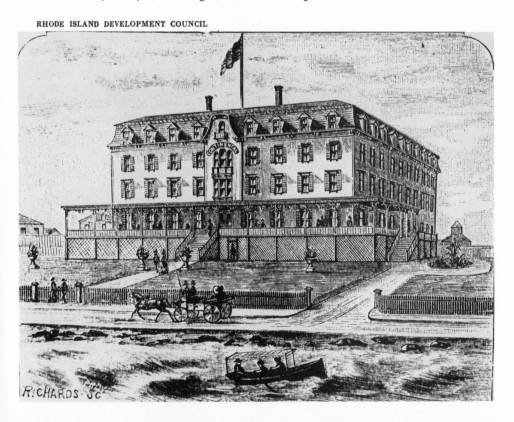

THE GREENBRIER

Portsmouth & Roanoke
RAIL ROAD.

Cars and Carriages, for the Transportation of Produce, Merchandize, and passengers, on the *Rail Road* between *Portsmouth* and *Suffolk*, are now running twice a day.

Southern society flooded to White Sulphur Springs both before and after the Civil War. The manager was so arbitrary he was known as the "Metternich of the Mountains." Foreigners were confounded to see the new republic, founded on equality, displaying social elitism. The most beautiful girls of the Southland, said one visitor, held "their court of love at this fountain." This early train depicted in a Southern poster was little more than a stagecoach running on rails.

Bill of fare on the steamer *United States* offered a full meal for $1. That included roast beef, turkey, goose, veal and lamb as well as bowled fowls, a choice of seven puddings, ice creams and nuts. To wash it down, there was Heidseick (sic) or Roederer at $3.50 a quart or Guinness' Dublin Porter at 30 cents a glass. Railroad timetable of 1879, below, guaranteed safe passage through hostile Indian country and Black Hills storms.

Wine List.

CHAMPAGNE.

MARCEAUX,	(quarts)	$3 00
"	(pints)	1 75
VERZENAY,	(quarts)	3 00
"	(pints)	1 75
VERZENAY, (Dry)	(quarts)	3 50
" (do.)	(pints)	2 00
VERZENAY, Green Seal,	(quarts)	3 50
"	(pints)	2 00
ROEDERER, (Carte Blanche,)	(quarts)	3 50
"	(pints)	2 00
" (Dry Sillery.)	(quarts)	3 00
"	(pints)	1 75
HEIDSEICK,	(quarts)	4 00
"	(pints)	2 00

CLARET.

PAUILLAC,	(quarts)	1 50
	(pints)	75

SHERRYS, &c.

SHERRY WINE, (Harmony,)	2 50
PORT WINE, (London Dock.)	3 00
HOCK,	2 50
JOSEPH'S HOFFER, (Rhine Wine)	2 50

ALE & PORTER.

MUIR'S Scotch Ale,	50
YOUNGER & SONS' Ale,	50
PHILADELPHIA,	35
DOW'S PALE INDIA,	30
GUINNESS' Dublin Porter,	30
CARPENTER'S Champagne Cider,	30

Waiters are supplied with Cards.

Shellsburg, Iowa 1877

THE ONLY ROUTE TO THE BLACK HILLS OPEN THE YEAR ROUND UNION PACIFIC R'Y and SIDNEY STAGE LINE, AVOIDING BAD LANDS, HOSTILE INDIANS AND THE DANGEROUS STORM BELTS OF THE NORTH.

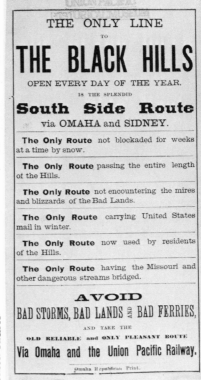

THE ONLY LINE TO

THE BLACK HILLS

OPEN EVERY DAY OF THE YEAR.

IS THE SPLENDID

South Side Route

via OMAHA and SIDNEY.

The Only Route not blockaded for weeks at a time by snow.

The Only Route passing the entire length of the Hills.

The Only Route not encountering the mires and blizzards of the Bad Lands.

The Only Route carrying United States mail in winter.

The Only Route now used by residents of the Hills.

The Only Route having the Missouri and other dangerous streams bridged.

AVOID

BAD STORMS, BAD LANDS and BAD FERRIES,

AND TAKE THE

OLD RELIABLE and ONLY PLEASANT ROUTE

Via Omaha and the Union Pacific Railway.

Omaha Republican Print.

Potter Palmer, a Chicago millionaire, opened the Palmer House in 1873. With 700 rooms, it was called the "Marvel of the Midwest." Built only two years after Mrs. O'Leary's cow kicked over the lantern, the hotel was said to be so fireproof that Palmer invited other hoteliers to build a fire in any of his apartments if he could do the same in theirs. Said to contain the costliest art ever used to ornament a building in the United States, it also had an enormous kitchen, pictured here, top. Journeying to take the waters at spas was popular in America and at the famed watering places abroad. This version of immersion in therapeutic waters was sketched at Interlaken, Switzerland, in 1863.

The pool at Miami's Royal Palm Hotel at the end of the century holds little resemblance to the splashy swimming holes that were to appear a few decades later. The bathing costumes were to change, too. St. Augustine and the Palm Beaches were developed long before Miami, which was still jungle when Henry Flagler's first passenger train arrived there in 1896. The village had a voter's register that numbered 502. Steamboat travel on the lakes and rivers of the United States was elegant, the decor fussy, as the interior of a salon on the steamer *Grand Republic*, bottom, shows. Some ships carried barber shops, and service and food was as good as in the best hotels. In crowded times there were three table sittings.

Once the railroad had linked the rails that made it possible to ride from New York to St. Augustine, Florida, where hotels had been built, the march began down the east coast. The steam train, above, of the Jacksonville, St. Augustine and Halifax Railway arrived in Daytona in the 1880s. Guests at the Royal Poinciana, below right, included Vanderbilts and Whitneys. In the West travelers were riding this plush sleeping car.

HAYNES INC., YELLOWSTONE PARK, WYOMING

IV | *Golden Years (1900–1914)*

IT WAS A RICH and bountiful era while it lasted. The ships that steamed across the Atlantic grew even longer and even faster. The grand hotels were so grand they were known as palaces. And itineraries were so exotic they were called grand tours.

But this was all frivolity before the storm. The era, short as it was, belonged to the mechanics tinkering in the machine shops, for while wealthy Americans went tripping off hither and thither, it was the engineers who stayed at home and developed the automobile and, as it was called then, the aeroplane, a grown-up toy that would change the ways of man's movement for all times.

By century's turn there were twenty-three thousand cars in the United States, very few highways, but lots of jokes.

He: What made you so late?

She: I came up in my automobile and passed here three times before I could get it to stop.

Jim: There were eight hundred killed in the Philippines.

Herman: I didn't know they had that many automobiles out there.

Even Mother Goose had a go:

> Mary, Mary quite contrary
> How does your motor go?
> With ringing bells and shocking smells
> And cylinders all in a row.

People who drove cars were called chauffeurs, or better yet

autoneers. Out in Ohio, people collapsed in gales of guffaws over the difficulties of one autoneer who, encountering a horse and carriage, brought his machine to a halt and jumped down to help lead the horse past it. A small boy who held the horse's reins was calm, but the lady beside him was screaming. "Let me take care of the horse," the boy shouted. "You just lead Maw past that contraption."

The marshal of Sparta, Ohio, dispatched a written warning to one local auto owner. "Not being able to see you personally," the lawman wrote, "I take this method to warn you to stop at once speeding your auto through this town as you endanger the lives of citizens. You are not to run it above five miles an hour under penalty of arrest and heavy fine."

The history of the speed traps began almost with the emergence of the automobile. Constables disguised themselves as road workmen, stood ready to swing a rope across the highway to stop any motorist exceeding the speed limit of eight miles an hour. In Lancaster, Massachusetts, the cops posted themselves at the bottom of a hill, a point at which autoneers had to gather all speed possible in order to get their machines over the next hill. They caught eight who were brought into court and relieved of fines of $15 each.

Lawmakers and the police were goaded by anti-automobile societies which lobbied for stricter limitations. One bill called for any self-propelled vehicle to come to a complete halt upon approaching any crossroads. Said the statute:

> The engineer must thoroughly examine the roadway ahead, and sound his horn vigorously. Then haloo loudly or ring a gong. Afterwards, he must fire a gun of sufficiently audible character to be heard at a great distance. Thereupon he will dismount and discharge a Roman candle, Vesuvius bomb, or some other explosive device as final warning of his approach.

Any autoneer, chauffeur or engineer who followed the letter of this law could, of course, be arrested at the same time for disturbing the peace, breaking the Sabbath, illegal transportation of explosives and discharging firearms within the city limits. The politicians, not yet sensitive to the feelings of the auto workers, much less the auto owners, strove to protect the horse. Once they had a law which held the driver of a car which had caused a team of horses to bolt and rear to be liable to a fine of $100. Should the horse refuse to pass the car at all, the driver was required then to camouflage his car with a cloth painted to look like a landscape.

Not quite so prophetic as it was destined to become in later years, the New York *Times* opined in an editorial, "Americans will never learn to love the mechanical wagon, because they will never get used to speeding along the road behind nothing." Equally cloudy in its evaluation of human capability, a medical journal peered carefully into the future and predicted, "If the machine ever attains the unlikely speed of 80 miles an hour, it will have to drive itself, for the human brain will be incapable of controlling it." Another moralistic paper noted, "Small wonder that ladies of the ensemble and other flashy theatrical types prefer this type of vehicle. Its machinery is activated by material which burns not unlike brimstone."

The ladies of the ensemble not only preferred this type of locomotion, they were even singing about it. Virginia Earle, in *The Belle of Bohemia,* a success of 1903, warbled one of the first car songs, "My Mobile Gal." Later came "In My Merry Oldsmobile" which was such a resounding success that every car company tried for a similar hit. "Ray and His Little Chevrolet," "The Little Ford Rambled Right Along" and "On the Back Seat of the Old Henry Ford" were all song sheets of the day, but none topped Gus Edwards' efforts for Oldsmobile.

As for the early automobile admen, they turned to the newspapers and magazines and came up with such sales talk as:

"Climbs Hills Like a Squirrel and Eats Up the Road Like an Express Train."

"Hills and sand become level land."

"A common sense car with no tender or delicate parts."

Or, as a knock to the competition:

"Hot or cold weather are demoralizing to the horse."

Some of the hardships that demoralized the car owners were not mentioned, at least in the ads. Automobiles began bumping into one another. Mr. Philip Hagel earned the unhappy honor of having to pay the first automobile damages when he drove his French De Dion Bouton limousine up Manhattan's Broadway. This momentous ride cost him $48,000, the amount being the total of damage judgments awarded owners of New York horses frightened by their first sight of a horseless carriage.

The initial automobile death occurred in 1900 and the victim was Hieronymus Mueller. While filling the gasoline tank of his automobile, he came too close to a match which ignited the gasoline and set his clothes on fire. Said the insurance adjuster with some premonition, "Mr. Mueller was so badly burned that he died in a few hours. This is the first fatal case of this kind which has come to our notice. There are undoubtedly more to follow although the manner of accident may differ."

Robert King of Pittsburgh had the unhappy honor of being involved in the world's first automotive collision. As he described his experience,

> I was returning from a drive one evening in 1900. As I neared our home on Negley Avenue, my wheels became wedged between some planks laid on either side and between the streetcar tracks while some grating work was in process. I saw the headlight of an approaching trolley car. I leaped out, waving and yelling, but the car came on. This being the end of the line and no passengers being aboard, the motorman had left the controls and gone to the rear of the car to talk with the conductor. My beautiful Riker was smashed. I sued. In court the attorney for the streetcar company told the jury that the company which provided transportation so that the poor working man could ride home from his work when he was too weary to walk should not be penalized for damaging a rich man's plaything, which was good only to frighten women, children and horses. The jury awarded me $1.00.

Motorists in this era bought their gasoline at a hardware store and then filtered it through a chamois cloth. They carried enough equipment to replace a broken fan belt, change spark plugs, insert new gaskets and pump and patch tires, all considered minor repairs. Anything more ambitious required the attention of a blacksmith. The road system was a nightmare and heaven help the highwayman when it rained. Hopelessly mired motorists who had to be hauled out of the mud provided a new source of income for the farmer and his horses. Some farmers were supposed to have rushed back to the roads in the dark of night to refill the larger mudholes with water.

There were other automotive frauds, too. The Travelers Insurance Company still keeps this gem of a report:

> One day a tester reported he had killed a dog, which had shot out of some bushes and directly in front of his car. A woman soon appeared, demanding $10 for the death of her pet. The Travelers paid her. About a week later, the same thing occurred at the same spot, and the same woman appeared for another $10. When, about two weeks later, this occurred once more, we called in a detective.
>
> He got the facts: this woman had been buying dogs at the dog pound for a dollar each. She had two young sons. When a car of ours approached, one son would hold a dog in the bushes on one side of the road, while the other, holding a bone, was hiding in the bushes across the way. At the right moment, Number Two Son waved the bone, Number One Son released the dog, and the animal sprinted across the road.

In addition to all these operative difficulties, the owners of the first cars even encountered social criticism. Automobiles were called "rich men's toys" and worse. One glib social orator of the day denounced them roundly as "Devil wagons in which our wealthy and flanneled *samurai* are hurtling down the primrose path of privilege at twenty miles an hour. . . ." When he was still president of Princeton University, no less an intellect than Woodrow Wilson was moved to blame the automobile, of all things, for certain creeping political manifestations. "Nothing has spread Socialistic feeling more than the automobile," he said. "They are a picture of arrogance and wealth with all its independence and carelessness." Some nasty rumor-mongers whispered the word that during the fire in San Francisco, cars were used to rescue the aristocrats from their lofty perch atop Nob Hill. It was noted that autos weren't even an American invention and that, of course, made them suspect anyway.

Despite the snide references to the rich, it was they who were truly responsible for popularizing cars and their side product, tours. Car owners first banded together in New York in June of 1899 and formed an association. The club moved, first of all, to eliminate the restrictions against motor vehicles in New York's Central Park. Two of the wealthiest members cranked up their engines and drove into the park on a dare. They were not arrested, and automobilists all over the country took heart. By 1902, the American Automobile Association was formed. One of its aims was to erase prejudice on a national basis. There was certainly plenty of it around. Cars were categorically barred from national parks, but then so, for that matter, were hikers. Any seeker after pristine pleasure was required to take a horse-drawn sight-seeing wagon. The automobile ban in parks remained in force until 1915.

There being neither a war nor an income tax, the public conversation and the public pleasure turned toward cars. More and more people bought them every year, venturing away from the hearth on weekend jaunts, carrying with them, besides all that emergency equipment, a copy of the *Automobile Blue Book*. The *Blue Book* carried maps, but just to be on the safe side, it also described each suggested route. "Turn right at red barn, then 1.5 miles to P.O. and a General Store. Take left fork on narrow dirt road. Maybe bad in wet weather . . ." was a typical destination description.

The first daredevil who embarked on a cross-country trip— it was 1903—took 52 days to reach the West Coast. The trouble was that there was really not very much road to chart, for by 1904, only 7 percent of the 2 million miles of publicly owned U.S. road could be called improved. And "improved" meant only that the stumps had been removed, that the pathway was relatively flat, and that the grass growth would not entangle a car's undercarriage. There were only 150,000 miles of surfaced road. And "surfaced" meant that the highway had a coating of some sort—either oyster shells or sawdust or rough planks. In all the country, exactly 144 miles of road were paved with macadam.

A Bostonian named Charles Glidden, who had done well in the telephone business and who was an autoneer and a pioneer tourist, donated an elaborate trophy which the American Automobile Association was to present to the winner of a reliability tour. Participants were to journey from New York to Bretton Woods, New Hampshire. The tours were to show the glories of touring the land, and at the same time to emphasize the deplorable condition of the nation's roads. The first trip in 1905 drew such aristocratic entrants as Ransom Olds himself, Percy Pierce of the George and Pierce Co. and Walter White of White Motors.

The first Vanderbilt Cup Races, the year before, drew a crowd of twenty-five thousand. They proved such an unmanageable horde that the race finally had to be stopped. The trophy went to a French Panhard, which averaged fifty-two miles an hour around the track. Most of the participants were wealthy amateurs. It was just such people with the money to buy and maintain expensive cars who set off the trend to touring. They went probing through the countryside, keeping a keen eye out for those twin bugaboos of early motoring, bad weather and bad roads. Soon the hotelmen of New England, a group never exactly unmindful of new ways to turn a dollar, were sending out tour books and brochures. So many autoneers were pulling up at hotels in the Northeast that innkeepers had to establish some rules for decorum. In Maine, the Poland Spring House posted an ordinance prohibiting the wearing of dusters and goggles in the lobby on Sundays. Ladies turned to their magazines for advice. "I have been invited for a week's trip in an automobile," said an inquiry received by the *Ladies' Home Journal*. "I feel that a motoring bonnet which ties under the chin is unbecoming. What do you suggest?" The *Journal* told the lady to deck herself out in a blue taffeta tam-o'-shanter.

"See America First" proved a versatile slogan. While it was patriotic, it also sold cars. By 1906 there were seventy-five different makes to choose from, and a number of different ways to make them run. The list of names ranged alphabetically from the Albany and the Apperson to the Success and the Zimmerman. There was a Packard as well as a Pickard; a Pierce Arrow as well as a Sharp Arrow. You could ride in a Klink or a Coyote or pick either a Buggycar or the Bugmobile. (Shades of Volkswagens to come.) The Stanley Steamer, which established a sensational record of 27 miles an hour, was known as the "Flying Teakettle." It took a strong early lead over what was called then the "internal explosion" machine. By 1912 there were dozens of makes of cars, some propelled by compressed air, some by electricity. "Travel in white gloves, go electric," the slogan said. "No fuss, no fumes, quiet as a pussycat's purr." It wasn't long before the competition answered, "Why risk electrocution?" From the opposite bench came the response: "The machine that steams is the car of your dreams."

When it came to accessories, the autoneer was faced with a myriad of new inventions, not the least of which was Bosco's Collapsible Rubber Driver. This device, when inflated, looked more or less like a man. Placed at the wheel, it would scare away would-be thieves. "No thief ever attempted to steal a car with a man at the wheel," the ad for Bosco's Collapsible Driver said. "It's so lifelike and terrifying that nobody a foot away can tell it isn't a real live man." When not acting as a sort of pneumatic scarecrow, Bosco's rubber man could be stored flat under the car seat.

The social notes filtering down from Newport as early as 1906 recorded that the members of the colony in summer residence there were rolling along Ocean Drive in horseless carriages. Two years later, *Vogue* took notice that "certain circles," as it called the select group to whom it looked for leadership, were sidestepping all the frenzy of rail travel, and journeying from place to place in their new cars. The King of the Belgians, who frequently shuttled back and forth between Paris and Brussels, was now making his trips in an automobile. Said *Vogue*, "Is it not a truly royal way of getting about . . . ?"

Americans and Europeans didn't have to be told twice. As early as 1906, the first of a string of books began appearing, describing the high adventure of motoring through Europe.

Winthrop Scarritt, author of *Three Men in a Car*, published that year, told of the terrific speed with which cars are driven in Paris. "It is not an uncommon occurrence" he wrote, "to see cars driven up the Champs-Élysées at the rate of 25 miles an hour, and occasionally under the very nose of the police, one is seen going at the rate of 30 to 35 miles an hour. The roads were so good in France that one was tempted to push down the pedal and unseeingly go roaring through towns laden with ancient Cathedrals and priceless works of art at 20 miles an hour." The cost of car travel came to about $12 a day, including food, lodging, gasoline and the services of a chauffeur, who, out of his fee of 20 to 25 francs, was expected to pay for his own expenses.

Frank Presbrey, who wrote *Motoring Abroad in* 1908, found garage facilities in almost every hotel, where storage, washing and "brassing" would come to 2 francs. Gasoline cost about 50 cents for two gallons and it came in sealed tins. Tires, horns and sirens were 50 percent cheaper than in the United States. Among the tips passed along by early car travelers: Put a padlock on the hood of the car because French mechanics are so curious about foreign makes that they often tinker with the mechanism; while French roads are probably the best in the world, one must watch out for heavy nails which peasants wear in the soles of their wooden shoes. The motorists could count on a puncture a day unless he attached a protective shield to his wheels.

The big steamship companies found a profitable sideline in shipping cars abroad. In Paris, Thos. Cook and Son sent its sightseers out on tours in charabancs. Even Edith Wharton wrote a book about touring, in which she said, "The motorcar has restored the romance of travel. Freeing us from all the compulsions and contacts of the railway, the bondage of fixed hours and the beaten track, the approach to each town through the area of ugliest desolation created by the railway itself, it has given us back of the wonder, the adventure and the novelty which enlivened the way of our posting grandparents."

Edith Wharton was so charmed by her tour by automobile that she called her 1909 book *The Motor Flight Through France*. It had, of course, nothing at all to do with flying, but the freedom of travel which car travel permitted made it seem that way, and Mrs. Wharton was, in fact, signaling the commencement of an enormous trend that would send Americans rolling all over their own country and in foreign lands as well. The advent of the automobile seemed also to spark a national mood for leisure. Men, and also women, of all things, took up tennis, croquet and even golf. One town, realizing the possibilities, laid out an eighteen-hole course in a pasture and put up a sign. "Accessible to everyone . . . rolling green fields over which you can ramble and hit things. Golfing is healthier than the tango . . . and easier. Anyone who ever killed snakes can do it. . . ."

Driving to the seashore became a double pleasure. The Sears catalogue of 1905 offered an advertisement for a lady's bathing suit with attached bloomers made of brilliantine. "Has large sailor collar trimmed with two rows of white cord, one row braid," the ad read. "Sleeves trimmed to correspond. Detachable skirt, waistband trimmed with rows of cord and a row of braid, trimmed around bottom to correspond. Colors black or navy with white trimming . . . $2.98."

While they were permitted to get onto the tennis courts and the golf links too, provided, of course, their skirts covered their ankles, ladies were expected to dress with decorum on the beaches. One who appeared in Atlantic City in a bathing suit considered too short by the populace gathered there for the summer of 1913 was actually assaulted by an outraged mob. "Outright killing doesn't seem to have been the purpose of the offended throng," editorialized the New York *Times*, "but they frightened the too close imitator of the Naiads something more than half to death, by belting her with epithets and sand, and what they would have done had it not been for the rescuing police is unknown. . . ."

Despite the conservative views of Atlantic City, the Jersey coast was considered the most democratic of all American resorts. It became, in effect, an all-year-round playground, an idea that was probably originated by Brighton in England. In summer, though, the beaches from Bar Harbor—which had just elected to permit automobiles—down to Cape May seemed as congested as the cities. The summer houses and summer hotels

stretched for fifty miles southward from Sandy Hook along the
Jersey coast almost in one continuous streak. For the first time,
Newport's preeminence in the social saltwater world was challenged by Southampton, the Long Island resort 100 miles from
Manhattan, a brisk, gay watering hole brimming with high
times and high fashion. Southampton thrived on parties and
what made its party successful was its seemingly endless pool
of eligible attractive men, all of whom came down from the
nearby big city. Bar Harbor, in the remote Northeast, was
always supplied with men over seventy. Newport, way up
there in Rhode Island, was depending upon boys of twenty.
Said one social observer of the day, "Proximity to the great
cities makes it easier to get men, who, like fresh vegetables, are
always easier to find in town than in the country."

Niagara Falls, after holding its title as the nation's prime
vacationland for uncountable decades, was, at last, *déclassé*.
Canada was discovered as new domain for explorers. Travelers
from the States pushed across the borders into Nova Scotia, on
to Cape Breton Island and even into Newfoundland. Virginia
was busy once more, building elaborate hotels near the springs.
Asheville was popular with anyone suffering from asthma, the
early stages of phthisis, hay fever or even nervous prostration.
It only got up to 39 degrees in winter, but July and August,
which were cool, and February and March were the big seasons. Then, the Battery Park, which cost $4 a night, and the
Kenilworth Inn which charged $3, were heavily booked. The
Kenilworth, two miles from town, was especially the favorite
of visitors who had come to see Biltmore, George Vanderbilt's
fabulous $4 million mansion which had already been aban-
doned by the family, and had been opened to the public as a
tourist attraction.

Despite the decline of Niagara Falls, Saratoga, also in New
York, stayed in its glory, its giant hotels giving splendid shelter
to 20,000 visitors during the racing season. The Grand Union
possessed 2400 feet of street front and 1500 beds, which were
available at $4 each. Its brochures pointed out that the dining
room of the hotel was 275 feet long.

The Borscht Belt was being born, even though in its early
years restrictive policies kept most of the Jews away. In this

pre-Grossinger's era, the Hotel Kaaterskill at the cool 2400-foot level was the ranking establishment in the area. It had no fewer than 1200 rooms to rent and it had little trouble filling them at $25 a week.

One could see the Catskills from the Shawangunk Mountains, where in the nineteenth century the Smiley Brothers, Quakers from Maine, had built an incredible castle called Mohonk Mountain House. The newest part was added in 1902, bringing its capacity to 600. It was usually filled, too, even though no liquor was served in those days and there were strict rules about smoking.

California was just beginning to earn a substantial reputation. More than 1.5 million people lived there, and their picture postcards sent back east began to attract visitors. San Francisco's racy reputation certainly encouraged the gay spirits to journey west. Commenting on the strange fact that a few warehouses full of spirits had escaped the fire, a popular doggerel of the day inquired,

> If, as they say, God razed the town for being over-frisky
> Then why did He burn the churches down, and leave
>   Hotaling's Whiskey?

The frostbitten of the Northeast continued to journey south to Florida. Rail connections were good and it was possible to get from New York to Jacksonville in twenty-six hours, for $26.30 plus $6.50 for the sleeper. Sailing aboard the steamer, which left three times weekly from pier 36 in the North River, proved a more leisurely way to travel. The Clive Steamship Line ship took two and a half to three days to reach Jacksonville and charged its passengers $25 for the voyage.

Perhaps the most unusual resort of the era was Ocean Grove on the New Jersey coast. It was cheap; the best hotels charged no more than $3.50. It drew at least twenty-five thousand people a year. Why they came is hard to guess. The resort was established in 1870 by an association of the Methodist Episcopal Church. It was a religious resort, an autocratic spa at which people voluntarily elected to spend their summer vacations under conditions that might be considered distressing. The drinking of alcoholic beverages and the sale of tobacco were both prohibited. The gates were closed at 10 P.M. daily and all day on Sunday. Bathing, riding and driving were not permitted on Sunday. No theatrical performances of any kind were allowed. The vacationers attended religious meetings which were held daily. The meeting hall could hold ten thousand and was dominated by an organ that seemed powerful enough to spring the rafters with its hymns. For other divertissement visitors could gaze at a model of the city of Jerusalem spread out under a nearby tent. Neighboring Asbury Park undoubtedly owes its rapid growth as a summer resort to the many visitors to Ocean Grove who thought something had been left out of their vacation and headed to more happy if heathen surroundings.

Over on the Brooklyn Riviera, Coney Island was not the tawdry popcorn amusement park area which it later became, but one of the nation's leading seaside resorts. Luna Park with its six hundred thousand electric lights and its fabulous Dreamland was one of the latter-day wonders of the world. A twentieth-century Colossus of Rhodes shining in a glory of lights. The hotels along the seashore were done in grand style with broad lawns and bathing accommodations for hundreds of visitors. The beaches throbbed with electricity and exploding fireworks every night.

The trip to Coney Island could be done by elevated train or trolley car. Taxicabs were almost as scarce in those days as they seem to be during a Manhattan rainstorm today and arguments about fares were frequent. All guidebooks to New York advised that in case of a disagreement, the traveler was to take the hack driver summarily to the nearest police officer or to City Hall, where a full complaint could be brought to the mayor's marshal in room one.

Sight-seeing automobiles in New York City operated tours from the Flatiron Building several times a day including Sunday. The uptown trip cruised Fifth Avenue, Central Park, Grant's Tomb and Riverside Drive, took two and a half hours and cost a dollar.

Visitors seldom roamed completely throughout New York. They usually came into one section and stayed most of their time in this one small area. Businessmen, for instance, usually went to hotels closest to the budding business district, hardly farther uptown than Fourteenth Street. The Astor House at 225 Broadway was particularly popular with commercial gentlemen. Rooms started at $1.50 per night. Tourists of certain means headed toward the area between Madison Square and Central Park, staying at the Waldorf-Astoria, the St. Regis and the Gotham. Rooms were about $4 a night, often with meals included.

An increasing number of Americans found themselves with both the cash and the leisure time to venture around the world. The matter of money was not as serious as the matter of spare time. One globe-trotting diary writer records, "The world ticket from Lincoln, Nebraska to Lincoln, Nebraska costs $831." She left home with $1650, spent $1213, stayed away for five months and one week living "in a style strictly first-class." She spent another $250 abroad "to replenish her wardrobe and on souvenirs" and actually returned home with better than $200 in her pocket.

The most popular pathway began in San Francisco, continued to Hawaii, then crossed the sea to Japan, Hong Kong, Manila, Calcutta, Colombo, up to Bombay, to Cairo with a quick tour through Europe tossed in as the *pièce de résistance*. This pattern was varied occasionally with stops in China, usually at Canton, Shanghai or Hangkow.

Most of these Americans came home with the overriding impression that Asia was a continent of filth, disease and squalor. The child marriages of the East filled them with particular horror, and they wrote repeatedly of it in their diaries and travel books.

The ricksha disturbed them, too. The thought of a robust American being pulled through town by a frail, consumptive Oriental filled them with nagging guilts.

As these tyro travelers became hardened Asia voyagers, the idea became less disagreeable. "You discover immediately that the rickshaws in China are far more comfortable than in Japan," one traveler wrote. "They are more roomy and they are hung lower, and they roll smooth and noiselessly along on pneumatic tires drawn by swift-footed Chinamen, clad in blue denim uniforms. . . ."

The mercenary merchants of Japan were relentless in their pursuit of that new windfall, the American tourist. One shopper complained,

> You are beset by shopkeepers everywhere who send representatives to your hotel to ask your patronage. You find cards galore and envelopes filled with the most artistic and tempting advertising matter under the door of your hotel room. The best hotels no longer permit merchants to bring goods to the private rooms of guests, but you find them lurking in corridors and halls and in hotel parlors eager to display their bargains. When you go to the shops, you are met by bowing and salaaming clerks and proprietors who confuse you with a multitude of lovely things. . . . In some of the elect shops to which your guide conducts you, tea and cakes are passed about on lacquered trays by little Japanese girls, and you are treated as if you are an invited guest instead of just a shopper. I suspect that many a visitor is hypnotized into purchasing by this charming and polite custom of these shrewd little merchants of the Orient.

Hong Kong's tailoring talents had early beginnings which were equally appreciated. "Never mind the heat," one voyager wrote, "you may order a pongee or linen suit made for yourself, your husband or your daughter, and it will be delivered on board the next morning."

Sharp trading practices were not confined solely to the Orientals. A crafty Occidental confided, "The best time to barter and bargain and buy to your heart's content . . . is just before the ship sails—immediately after the natives are warned off the deck by the ship's officers. This is the crucial moment. It is a case of now-or-never, and the prices fall like magic."

Crossing the long stretches of ocean to reach these exotic bargain-decked outposts consumed a lot of time, and the passengers on shipboard, soon tiring of the games and shipboard activities, conspired to fill their spare moments. Often the passengers would form clubs, meeting before and after each port of call. The prestop meeting was to learn as much as possible about what they were going to see; the poststop meeting was for each person to tell what he saw. The games became more ferociously contested the longer the ship was at sea. The ladies conducted potato races, needle-threading contests, nail-driving contests and hairdressing competitions on the deck and in the salons. The younger men indulged in "spar-pillow contests." The two contestants sat opposite each other on a ship's spar about five feet in circumference and seventeen feet long erected over the deck, which had been suitably padded. They then threw pillows at each other until one or the other was knocked off the spar. Crossing the equator was a particularly rowdy affair as the "shellbacks" initiated the "pollywogs" into King Neptune's kingdom by painting faces and ducking people into pools of water.

Proving that almost anyone, even ballplayers, could gain something out of a foreign trip, the New York Giants and the Chicago White Sox, both major-league ball clubs, took off on a global tour which lasted from October 1913 to March of 1914. The game they introduced in Japan proved an immediate fascination and was quickly adopted.

More than ever before Americans seemed now to enjoy transatlantic travel. The $50 fare, provided one didn't mind going steerage, was available on all lines. The crossing took six to nine days. The most beloved and luxurious ships of the era were the four-stacked *Lusitania* and the *Mauretania*, both 38,000 tons and 785 feet long. They jointly held the Atlantic Blue Ribbon for the fastest crossings.

It was true that transatlantic travel did take a dismal dip in 1912 when the *Titanic* rammed an iceberg and went down with more than 1500 passengers, but the tragedy did little to keep Americans from traveling abroad in the years that followed. New ships came into service, each boasting more extravagant conveniences—cafés, swimming pools, garden lounges—and their lure was irresistible.

Other popular ships flew under the banners of the North German Lloyd and Hamburg-America Lines, the French Line, the Red Star Line and the White Star Line. The *Vaterland*, when it sailed, was a new giant, almost 1000 feet in length. She had room for 5000 passengers and crew, but made only two transatlantic crossings before she was seized by the Allies and became the troopship *Leviathan*.

Carrying their depleted bottles of Mothersill's Seasick Remedy ("Be happy and well while traveling") many of the American tourists debarked from their ships after a week at sea and headed straight for the American Express Company offices in Paris. The role was new for American Express, which had not expected to participate in the travel business beyond selling traveler's checks. The turning point in the company's history, and fortunes, came in 1894, when Mr. William Swift Dalliba was sent to Europe by J. C. Fargo, president of American Express, to develop westbound freight shipments from the continent. Dalliba, who had once been an officer of the law in Caribou, Idaho, but now affected a morning coat and shining top hat, plunged into the affairs of the American community in Paris, and eventually became its dean.

Under Dalliba's guidance American Express opened its first foreign office in Paris in 1895, followed by other offices in London, Southampton, Le Havre, Liverpool, Hamburg and Bremen. In 1901 the most famous American Express office in the world opened at 11 Rue Scribe in Paris. It was to be a nest to which American birds of passage would come flapping for years to come. The flatiron-shaped building still stands in the triangle where the Rue Scribe and the Rue Auber converge on the Place Charles Garnier beside the Opéra. At first, American Express quarters occupied only the front piece of this wedge, yet even then the office was imposing with its four tall Corinthian columns supporting an elaborately decorated ceiling, and its branch stairways with their graceful wrought-iron balustrades lifting up toward the mezzanine.

As soon as the new office opened, French predators decided that a good deal of American money was concentrated here for the taking, a popular French concept of some antiquity. One night the safe was ripped open by dynamite, but the thief, a desperado named Eddie Guerin, was captured and sent to Devils Island.

Although most Americans on arriving in Paris headed immediately for American Express, the firm insisted that it was not in the travel business. President Fargo thundered in a letter, "I will not have gangs of tippers taking off in charabancs from in front of our offices the way they do from Cook's . . . we will cash their traveler's checks and give them free advice . . . that's all." Famous last words.

Soon American Express employees were securing railroad tickets, making hotel reservations, planning itineraries and finding lost baggage throughout the whole of Europe. Travelers waving good-bye to their friends on the New York piers gave them a simple European address: c/o American Express Company, Paris. Dalliba found himself running a post office, a booking bureau, a bank, a travelers' aid society and a hand-holding establishment for nervous young ladies. For a long time, the company derived little direct profit from these services. American Express's mail room is still free today.

Fargo was not one to perceive trends in the dim haze of the future. "There is no profit in the tourist business as conducted by Thomas Cook and Son," he said, "and even if there were this company would not undertake it." But as he was thundering these edible words, Dalliba was writing from Paris saying that the ticket department was handling five thousand letters a day and needed more space to accommodate tourists. Fargo barked back, "Enlarge your organization when the express and freight business crowds it, but let the tourists crowd themselves."

By 1912 the transatlantic battles between the two offices of American Express ended. The company became in reality a travel agency acting as ticket agents for the London and South-western Railroads, and organizing a wide range of sight-seeing tours through Naples, Berlin, Paris and London. Indeed, charabancs crowded with tourists were leaving from the curb in front of American Express offices almost every day.

Dalliba hired Billy Dodsworth, a great friend of many notables of the day—Flo Ziegfeld, Maxine Elliott and John Drew—and they all patronized him at American Express when they came to Europe. It was Dodsworth's responsibility to take care of American tourists, and he developed a canny ability, or at least the reputation, for determining the credit rating of a customer on looks. If one looked solvent to Mr. Dodsworth, he had only to present his personal note in order to exchange an empty book of traveler's checks for a full one.

Travel to Europe never was so pleasant as in these two or three summers just before the outbreak of World War II. The great ships came in, unloaded their cargoes of touring Americans, and left loaded with tired sightseers on the way home. The *Lusitania*, the *Mauretania*, the *France*, the *Olympic*, the *Aquitania* and the *Imperator* were swift, opulent and abounding in every comfort and pleasure, crammed with visitors from the New World anxious once again to explore the delights of the Old. Marconi's marvelous new device, the radio, had dispelled the mystery and the loneliness of ocean travel. Europe was now only five days away from American shores, with new superliners conveying travelers abroad in satiny luxury that was far more elegant than one's own home ashore. No passports were needed anywhere in Europe except in Turkey and Russia. The opulent railroads such as the Simplon-Orient and the Train Bleu ran in luxury and on schedule. Newly perfected automobiles with shining brasswork and powerfully noisy engines rambled around Europe with hardly a breakdown. Close to 150,000 Americans were touring in Europe in this high fashion when in July of 1914 the Archduke Ferdinand was assassinated at Sarajevo.

Now all the Americans wanted to get home. There were not enough ships to take them, nor were there hotel rooms in Paris or London to hold them while they waited. None of the banks in panic-stricken Europe would cash their checks. Money had become worthless in the international conflagration. Newspaper headlines in the United States gave more attention to the plight of these stranded Americans than the news from battlefields and the death of thousands.

The U.S. Congress quickly appropriated $250,000 for relief of the stranded tourists and President Wilson sent two Navy ships to the Continent carrying $5 million in gold, made available by U.S. banks to their European branches, for disbursement to the travelers. The U.S. Government had not been so kind to tourists before or since.

One by one, refugees poured into London from the Continent. Mrs. Otto Kahn was forced to leave her two new shiny automobiles behind, but somehow managed to arrive in the

British capital with close to seventy trunks trailing behind her. When he could not get train space Alfred McCormick tried unsuccessfully to purchase a private yacht to make good his escape. Alfred Vanderbilt had to borrow two shillings from a hotel employee in order to get himself a shave, and Cornelius Vanderbilt became a "midnight flit," leaving his European hotel without paying his bill. A number of news stories dealt with twelve American Indians who had to leave their Wild West show in Poland to flee for their lives. A group of Wellesley students worked their way back on a tanker.

Herbert Hoover volunteered to take charge of an American Citizens Committee in London which attempted to give aid to these refugees as they poured out of the battle areas toward safety. Hundreds of them were issued cards—possibly the first credit cards—which identified them as U.S. citizens without funds and encouraged commercial channels to allow these travelers credit to get home. In six busy weeks, the committee handed out more than $1.5 million in credit to the tourists and shipped 120,000 of them back to the United States. Ten vessels were chartered to handle this westbound traffic. Surprisingly, not every passenger was delighted to go aboard and be free of the threat of the war. A number wanted guarantees against possible submarine attack. One discriminating passenger went on a hunger strike when given a berth in steerage. The transfer was accomplished without a single casualty and with only a few cases of hurt pride. It would be three years before Americans would come again to Europe in droves, and this time they would be in uniform.

As the Americans poured out of Europe, they might have heard a new sound growing in the air behind them: the whir of airplane engines. Aviation was catching the fancy of Europe's generals, where it had almost failed to capture the affection of the world's civilians before. World war was to give flight its first real chance. In the carnage it grew from a toy to an industry, becoming with dazzling swiftness the most dramatic method of transportation in the world.

As far as American travelers were concerned, probably the most significant flight of this era—aside from the Wrights in 1903—took place in 1911 when Galbraith T. Rodgers, a motorcycle driver with less than 70 hours of flight training, took off from New York in a plane owned by a soft-drink manufacturer and named for the bubbly drink.

The plane was called the *Vin Fizz*. Followed by a three-car train transporting his mother, his mechanics, his wife and close to $5000 worth of spare parts, the pioneering pilot lumbered across the country in a series of short hops. He was seeking a $50,000 first prize offered by William Randolph Hearst for the first flyer to cross the country successfully. The family, it turned out, was as badly needed for solace as were the mechanics for repairs. The plane smashed into the ground nineteen times, and by the time it reached California, the pilot had his leg in a cast and only pieces of the original air frame remained.

Rodgers spent 82 hours in the air and covered 3220 miles. He arrived nineteen days too late to win first prize, but he captured the imagination and affection of the traveling public nevertheless. America had been crossed successfully (if that's the word for nineteen crashes) by air, and the era of transcontinental aviation had made a small, bumpy but definite beginning.

V | *Whoopee! (1919–1929)*

PROHIBITION HAD AN intoxicating effect on the pattern of American travel during the 1920s. It turned the decade following the war into a confusing era, but a well-traveled one. "On the first they gave wine away free and on the sixteenth we went dry forever," cheered a prohibitionist when, on January 6, 1920, Prohibition became the law of the land. Sputtering with pleasure over the great moral victory of Prohibition, William Jennings Bryan predicted: ". . . it will be impossible to organize a national bootleggers' association any more than a national pickpockets association." Even as he was making this prediction, a steady procession of fast, sleek speedboats roared out to a fleet of rum ships anchored just outside the U.S. territorial limits; saloon keepers transferred their places of business to cellar bistros and down-alley cafés. The price of scotch went to 30, then 60 cents. Wallets began to bulge with speakeasy admission cards and travel to a wet Canada increased noticeably.

By 1921 Canada had become not just a scenic wonderland but the headquarters of a new enterprise: keeping the smugglers stocked in liquor. One of the most famous smuggling gangs affected the dress of nuns and priests and got away with untold cases of contraband. It was all almost too easy until the day the "clerical" band blew a tire at the border and one of the priests, declining the aid of the customs man, tried to change the shoe by himself. Experiencing some difficulty and a little harried by the inspector at his elbow, he suddenly exploded.

"God damn the son of a bitch!" he said, whereupon the customs agent thought he had better make a careful examination of the car. In July 1924 another slick idea came to life. It was quite commonplace for ice blocks to be shipped from Ontario to the United States. Then one day the weather turned scorchingly hot, and unlabeled bottles of gin and alcohol began to clunk and smash in the sudden thaw.

If there were problems in bringing the liquor from Canada to America, there was little difficulty in bringing Americans to the liquor. Canada suddenly became one of the most frequently visited American travel destinations. The cabaret business boomed there in 1921, and performers received almost twice as much for Toronto and Montreal engagements as they did in New York City.

Bimini in the Bahamas, a 500-yard-wide isle just fifty miles from the Florida coast, is known now as a fishing center, but during Prohibition it was a handy harbor from which to export bottles of Barleycorn to the needy across the straits. A four-story, 105-room hotel was built there in 1920 designed as an old tippler's home. The customers were to be brought by ferry. A hurricane blew in the hotel in 1926, and the thirsty looked for new spas or learned how to concoct juniper-flavored spirits in their own bathtubs.

Bermuda also became a favorite destination, and a cartoon in the May 1922 issue of *Vanity Fair* shows a pair of tourists returning from the British island, trying to fascinate a customs officer so he would not notice the bottles of booze bulging in their pockets.

Hiding contraband liquor in clothing was a favorite trick of returning travelers. Women passengers particularly usually carried large coats and many walked gingerly down the gangplanks. It was the custom of the times to drop a bottle in each coat pocket and stuff flat pint flasks in girdles.

In foreign lands around the periphery of the United States, liquor imports soared noticeably. American tourists were willing to hop over the border just for a quick drink. From 1918 to 1922, Canada increased her British liquor imports by almost 600 percent, Mexico saw its liquor importation rise 800 percent, while the British West Indies experienced a 500 percent increase in liquor traffic. In the Bahamas, just ninety miles off the coast of the United States, importation of liquor rose from less than 1000 gallons to almost 386,000 gallons annually, and in Bermuda the totals went up from 1000 to more than 41,000.

Ontario's Liquor Commission reported that it had nearly 300,000 permits outstanding, about 55,000 of which were issued in the "temporary" class, presumably to American travelers slipping across the border for a nip.

Prohibition created a new type of evening entertainment for the American traveler. In New York City, some of the older restaurants closed down. Soon the only spots in town still doing business in the old way were the Folies Bergère, the Palais Royale, the Little Club, the Café de Paris, Shanley's, Healey's and Reisenweber's. A novelty called the nightclub started in the summer of 1921, beginning with the Club Deauville on East Fifty-ninth Street. The nightclub was a descendant of the corner saloon and it proliferated. The fact that they were quite illegal didn't prevent over 5000 from blossoming in Manhattan alone, most of them in the midtown area. Most of them were screened in some sort of way posing as a tearoom or a respectable private house. Admission cards were printed and the little peephole door through which the applicant's identity could be checked became a symbol of the age. The fashionable trade developed their own speakeasies, some of which were to emerge, when Prohibition finally blew over, as respectable restaurants. The early café society also found it in vogue to travel to Harlem, especially to the Cotton Club, Small's Paradise and Connie's Inn. The Cotton Club spawned fine talent, among them Cab Calloway and the famed Bricktop, who started there in the chorus line and went on to become the darling of European cafés.

As the Victorian codes in America dissolved in a brew of raw whiskey and high postwar spirits, women, too, found themselves not only enfranchised but emancipated. The Nineteenth Amendment gave women the right to vote, and once they had that they also arrogated to themselves a number of new privileges, including the right to roam. Women who had let their men sail off on adventurous excursions abroad while they waited primly at home now insisted on going along. In a few years they became the majority of tourists, frequently traveling off without need of escort, and often urging their husbands to take them off to far places. The Cunard Line reported that some 60 or 65 percent of their passengers were women and girls, traveling both alone and in groups.

As their habits changed, so did their clothes. Ladies of the flapper era bobbed their hair and concealed their natural curves both fore and aft, all on the somewhat questionable theory that this new style gave them more sex appeal. A Chicago restaurant, following the taste of the 1920s into the near fringes of the ridiculous, dressed its waitresses in plus fours, apparently the last word in racy attire if affected by a female form.

A series of headline stories which brought glamour to distant ports of call excited the imagination of Americans. The most newsworthy archaeological find of the century was the discovery of Tutankhamen's tomb in 1922. When the tomb doors were opened in an Egyptian valley by Howard Carter, whose expedition had been underwritten by Lord Carnarvon, a dazzling treasure was unearthed. Headline readers and potential travelers were fascinated by details of the "curse" that followed those who had disturbed the 3300-year slumber of the Egyptian ruler. Many vowed to see for themselves the richness of the artifacts that were taken from the Valley of the Kings near Luxor.

Incited by the string of newspaper stories, the first escorted American party ever to travel through the length of the African continent left Cairo in January of 1924 on a journey that had been arranged by Thomas Cook and Son as carefully as if it had been the first trip of a timid midwestern widow to Paris and London. The pilgrims traveled by railroad, where this modern convenience existed, and where it did not they followed native guides and went on safari in the best traditional manner of African melodrama, trekking through wild veldt and jungle. It became an annual event.

Minting money in the stock market, sending their sons and daughters to college in increasing numbers, living high and handsomely in the most prosperous era the country had yet enjoyed, Americans went to see what the rest of the world and, for that matter, the rest of the country, looked like. By May, Americans ready to set sail were already well padded with traveler's checks, and steamship tags tied on steamer trunks were already fluttering in the spring breezes. These new explorers were spawned of a nation that had driven westward for a century carrying itself farther and farther away from the old homeland in the Old World. They had kept their eye on the Mississippi, then on the western plains, then on the Rockies and finally on the Pacific coast, moving always into the eye of the setting sun. With the West won and established, they suddenly turned around and, rolling eastward on transcontinental trains over the wagon ruts dug long before by the prairie schooners, they set out to find out whence they and their forebears had come. They went about discovering Europe, some thought, with the same headlong energy they had shown when they forged westward to found their own country. Before World War I the Germans had been generally held to be the most irrepressible travelers, and no one had ever doubted that the English—the relative few who journeyed abroad—had traveled farther because the Empire, and its accompanying interests, both crown and commercial, stretched around the globe. But the numbers of Americans taking to the high seas were growing enormously and so were their mileage records. Already the first trippers were being augmented by those who had covered the well-known places and now were down at the travel agent's office looking for something new, something rare, something exciting.

In January of 1927 the first round-the-world cruise by way of the southern hemisphere set out from New York. The voyage included the South Sea Islands, Australia, New Guinea, East Africa, South Africa and South America and was so successful that the next year four cruises followed in its wake.

In 1928, the Department of Commerce reported that more than 437,000 Americans had sailed abroad. "The way to see America," it was said, "is to get a chair at the Café de la Paix."

A leisurely sea voyage abroad was suddenly very much in vogue. The great ships which thousands of Americans took to France—their favorite country—were the *Ile de France*, the *Mauretania*, the *Aquitania* and the *Majestic*. Reported the

New York *Times*, "English ships that for years have been content with food as uninteresting as the two soups, one thick and one thin, which masqueraded under different names each day are busily hiring out famous chefs and advertising their cuisine. Single cabins are thrown into suites and furnished to look like guest rooms in a private home. Sun rooms are plastered to imitate stone. Elevators take the place of companionways and decks become porches."

Several hundred thousand Americans every year made Europe a summer playland for Americans. Most were leaping over the infringements of Prohibition and shaking off the restraints of their stolid lives at home. "Our tourist expenditures alone in Europe since the war," Herbert Hoover glumly stated in criticizing the Allies for their failure to pay their war debts, "would enable them to take care of the entire amount."

But even beyond the pure unalloyed pleasure-seeker there was that new foreign traveler, the businessman. As American industry began to expand into foreign markets and set up subsidiary factories abroad, businessmen began to commute between their home offices and their overseas operations. Rotary International held its eighteenth convention in Ostend in June of 1927. The American Legion registered 20,000 delegates to its convention in Paris in September of the same year, with side trips arranged to Belleau Wood, Château Thierry and Verdun. The desire for imported goods sent buyers from America's most enterprising stores into the most remote corners of the earth. At one point a representative of Marshall Field in Chicago had set up a cottage industry in China turning out lace handkerchiefs for eventual sale to midwestern matrons. Travelers with allied interests began to make trips in groups: farmers, Gold Star Mothers, former servicemen who wanted to show their families the battlefields of France and Belgium on which they had fought in 1917 and 1918.

Unfortunately the serious-minded business traveler, even when you added him to the scholars and the culture-seekers abroad in the Old World, were vastly overbalanced by the zealous, spirited child of the 1920s who knew America was better, wiser, younger, stronger than that tired old world that stretched from Brighton to the Balkans. In *Dodsworth*,

Sinclair Lewis has a character say, "Europe? Rats! Dead as a doornail! Place for women and long-haired artists. . . . Only American loans that keep them from burying the corpse. All this art! More art in a good shiny sparkplug than in all the fat Venus de Milos they ever turned out."

Even Will Rogers had a knock. Said he in the *Saturday Evening Post*, "Europe has nothing to recommend it but its old age. . . . You take the Guides and the Grapes out of Europe and she is just a Sahara . . . as a pure educational proposition or pastime it ain't there."

It may seem odd to find Will Rogers, otherwise remembered so warmly, aligned with the earliest of the ugly Americans. To the Europeans the Americans were the worst kind of boorish guests. They were arrogant about the purchasing power of their dollars. They drank and got drunk on the boulevards, in the venerable cafés and in the fine restaurants. They chewed gum and careened about the Continent flaunting their disdain by pasting valueless European paper currency on their suitcases. And they complained mightily about the prices. Newspapers in America were quick to pick up the controversy. Said one,

> Tourists to Paris this year . . . may look forward to an awful trimming according to F. M. Reinmund of 100 Williams Street who arrived yesterday on the White Star Liner *Olympic*. He said that prices had increased on the average of 40 per cent in Paris, though not in France as a whole. "The cheapest room with a bath in the center of Paris is now about $7.00," he said, "and a single portion of grapefruit costs 80¢. Cocktails which six months ago cost 14¢ are now more than twice that figure. All of their prices have advanced in like proportion."

The resentment was not long in smoldering, but the Americans didn't seem to care. Samuel Untermyer told a *Times* reporter, "Envy, hatred and contempt for American tourists were boldly paraded in France. Despite this rebuke, and the more they insult us, the faster we continue to come and the more of our money we pour into their coffers."

"A bunch of American tourists were hissed and stoned yesterday in Paris," read one report from Will Rogers, "but not until they had finished buying."

The result of all this ill feeling, finally, was one of the most violent outbursts against Americans abroad ever witnessed on any shore. Embassy stonings and burnings had not come into style as yet, but during the summer of 1926 angry Parisians gathered into mobs at the Place de l'Opéra and stormed the sight-seeing buses, crowded with American tourists, leaving from the American Express Company's offices. The worst incident, on July 24, sent a mob of thousands against the police lines. The cordon broke and the travelers were routed from their buses. Reinforcements of police finally brought order.

In what was for him a filibuster, President Coolidge reminded Americans that they were guests abroad, and advised those Americans who were "a species of a bumptious nature" that the people of Europe were, after all, still smarting from the rigors of a long and brutal war. While he was sure the attitude didn't represent the feeling of most Europeans, any American who didn't like the situation abroad and who was unable to muster some compassion for the local people ought to come home.

By 1927 Americans had gotten wiser to the ways of Europe, and not all were traveling in super-deluxe style. Some, a little more adventurous, and others, abroad on a budget, began traveling second class. Fewer travelers hired automobiles, which were expensive. Some were now willing to pay $3 for a room without a bath, rather than $5 for one with a bath. Very few were adventurous enough to jump aboard one of the airplanes of Europe's growing network, a timidity that seemed to amuse the Europeans. Most Americans had their ironclad itineraries made out long in advance, and they followed the routing assiduously, traveling mostly by rail. In Paris they might be tempted into the nightclubs, and they attended concerts and operas because the ritual of cultural indoctrination dictated it. But language and disinterest kept most Americans away from the sharp satire and the farces on view in Berlin and the operettas of Vienna. The greatest pleasure Europe seemed to offer was sitting around hotel lobbies with other Americans whose acquaintance had just been made, comparing sight-seeing, ships and cities and commenting to each other on the supe-riority of all things American.

Other Americans traveled to Europe, and in particular Paris, not to visit but to live. In *Exile's Return*, Malcolm Cowley recalled, "There was one idea that was held in common by the older and younger inhabitants of [Greenwich] Village—the idea of salvation by exile. 'They do things better in Europe: Let's go there.'

> Few of these people had travelled much in America—they knew little of the country they were rejecting. But many of them felt that the United States offered "no satis-fying careers open to talent."
>
> The younger and richer intellectuals went streaming up the longest gangplank in the world. . . . "I'm going to Paris," they said at first, and then, "I'm going to the South of France . . . I'm sailing Wednesday—next month—as soon as I can scrape together money enough for a ticket. Good-by, so long," they said, "I'll meet you in Europe. I'll drink to your health in good red burgundy, I'll kiss all the girls for you. I'm sick of this country. I'm going abroad to write one good novel."
>
> And we ourselves, the newcomers to the Village, we were leaving it if we could. The long process of deracina-tion had reached its climax. School and college had up-rooted us in spirit; the War had physically uprooted us, carried us into strange countries and left us finally in the metropolis of the uprooted. Now even New York seemed too American, too close to home. On its river side, Green-wich Village is bounded by the French line pier.

Once abroad, Cowley noted,

> Those who came in search of values found material values instead—the pleasures were cheaper. . . . It happened that old Europe, the continent of immemorial standards, had lost them all: had only prices, which changed from country to country, from village to village, it seemed from

hour to hour. Tuesday at Hamburg you might order a banquet for 8¢ (or was it 5?); Thursday in Paris you might buy twenty cigarettes for the price of a week's lodging in Vienna. You might gamble in Munich for high stakes, win half the fortune of a Czechoslovakian profiteer, then, if you could not spend your winnings for champagne and Picasso, you might give them day after tomorrow to a beggar and not be thanked.

Inflation worked very well for the American traveler in Europe. Cowley drew the scene:

> It was October of 1922 and Germany was entering its wildest period of inflation. When we crossed the border, German marks were selling 800 for the dollar; they had fallen to a thousand at Munich, twelve hundred at Ratisbon; in Berlin next morning, a dollar would buy two thousand paper marks or an all-wool overcoat. In the station we met Harold Loeb, the publisher of *Broom*, and Mathew Josephson; together, they were editing the magazine at a monthly cost of I don't what and they didn't know how many marks or dollars. Art was a liquid product that flowed across international frontiers to find the lowest level of prices. For a salary of $100 a month in American currency, Josephson lived in a duplex apartment with two maids, riding lessons for his wife, dinners only in the most expensive restaurants, tips to the orchestras, pictures collected, charities to struggling German writers. . . .

But Paris was Mecca and nirvana. Virtually all of the great writers and artists spawned by the 1920s were nurtured along the boulevards and the quays. Of course, not every expatriate in Paris was another Hemingway, Dos Passos or F. Scott Fitzgerald. Samuel Putnam described it like this:

> There were some who, unable to bring themselves to return to the States without the means of a decent livelihood, became drifters, floating from one job to another where a knowledge of English might be in demand, pro-

vided they were able to procure the always difficult *carte de travail* or labor permits for foreigners. Not a few became bartenders and even pimps, racetrack and prize fight hangers-on and the like. The meeting place, when they were in funds, was Harry's Bar near the Place de l'Opéra, where they could encounter and mingle with the tourists from home. They seldom came to the Left Bank—indeed, they seemed rather to shun the company of American writers and artists—but formed a motley colony of their own along with the touts, pugs, bartenders, pimps, tourist guides, gamblers, confidence men and vendors of pornographic pictures who had swarmed over to France during the years immediately following the War. They all constituted a demimonde that was half French, half American, a sort of weird amalgam of Brooklyn and Montmartre. . . . The type of expatriate that I have been describing was, of course, the exception, although there were more of those than one might think.

Said Gertrude Stein, who had held court for the expatriate intellectuals, "Paris was where the twentieth century was." One American, Gerald Murphy, whose father owned the Mark Cross store in New York, and who, with his wife Sara, had fled America for Paris in 1921, remembered it this way: "There was a tension and an excitement in the air that was almost physical. Always a new exhibition, or a recital of the new music, or a Dadaist manifestation, or a costume ball in Montparnasse, or a premiere of a new play or ballet or one of Etienne de Beaumont's fantastic Soirées de Paris in Montmartre—and you'd go to each one and find everyone else there, too."

In the days before World War I the French Riviera had been famed as a winter resort. The English came to escape the frigid fog. The nobility came down from Poland and Russia in such numbers that the onion domes of Greek Orthodox churches sprang up both in the French and Italian resort towns. When spring came the hotels closed and Cannes and Nice were as dead in summer as Newport or Bar Harbor in January. Just who got the idea to open the Riviera in summer is lost to history. Some say it was some American Red Cross girls at war's end, others insist it was Gerald Murphy and his wife Sara. At any rate, the Sella family, who owned the lovely Hôtel du Cap at Antibes, was persuaded to try one summer season. The Sellas obliged, keeping on one cook, one waiter, and one hall maid.

Reminiscing about those days in a *New Yorker* profile some years ago, the Murphys recalled that it was Cole Porter who had first gotten them interested in Antibes. He had rented a villa there and when the Murphys were invited down as houseguests they found the air delightful, the water "that wonderful jade and amethyst color." They scraped the seaweed off the beach called Garoupe and had a marvelous time. Cole never came back, but the Murphys did, staying first at the Hôtel du Cap and later in their own house. Picasso and his wife, Olga, son Paolo and his mother all came down to visit the Murphys, and they liked it so much they took a villa too.

The Murphys bought a place near the Antibes lighthouse and redid it handsomely into what became famous in that place and in those times as Villa America. The Murphy friends included Dos Passos, the Archibald MacLeishes, the Robert Benchleys, and Scott and Zelda Fitzgerald. Most guests stayed across the road in an old donkey stable located in an orange grove and called the Ferme des Orangers. Benchley renamed it La Ferme Dérangée. Fitzgerald, writing *Tender Is the Night* a few years later, made Dick Diver's home a cross between the Murphy's Villa America and composer Samuel Barlow's place up in the crags at Eze.

The Murphys lived in Villa America for ten years before returning to the States, but long before the ten years were up the Riviera, so unknown as a summer watering place when they came, was now packed with summering Americans. They came not only to Antibes, where the Sellas were now doing a booming summer business, but also to Cannes, to Nice and especially to Juan-les-Pins, a resort between the two, originated by an American named Frank Jay Gould.

When World War II forced her home, the *New Yorker's* Janet Flanner looked back on the twenty years with perception and nostalgia:

> The 1920s were a period of nomadization for thousands of Americans. They came in droves to France to gaze on herbage more succulent than their senses or their stomachs had known at home. Perhaps because we had (we thought) helped France to win the war; perhaps because we had in our pockets the first and last of our easy money, perhaps because we are born travellers; perhaps because we were originally of mixed European stock, we turned to France where on the whole, most of our ancestors never came from, French blood being scarce in the United States.

So many Americans were traveling in France in the Twenties that dire warning flags were hoisted. America was conquering Paris. T. R. Ybarra, a seasoned correspondent of the day, wrote a long treatise on the subject in which he recalled the Place Vendôme, center of the foreigner's Paris, was being called "the most beautiful square in America." Two new hotels put up their names in garish English letters, Ybarra noted, and one was advertising "600 rooms—600 baths." "Think of it," said Ybarra, "in a city where only a few years ago a hotelkeeper swelled up like a bullfrog when he could boast his place had 100 rooms and 20 baths." He found a hand-wringing expatriate American who rued the transgressions made by his own country's customs and quoted him as saying, "Why, I solemnly assure you, old man, at these new hotels they actually page the guests! Yes, sir—a nasty little page all covered with nasty little bright buttons goes about bawling out your name! You might as well be in Chicago!" A sign was said to have been placed in the window of a Place Vendôme shop reading "French spoken here." Soda fountains were cropping up. There was even an ugly rumor that the French were beginning to take less than two hours for lunch. Even President Poincaré was warning Frenchmen that if they would keep up with the pace of the times they had better tarry less at the table. Some thought the French would soon be eating at quick lunch counters or even stand up to munch sandwiches. But the French philosophy seemed secure in an unanswerable French riposte: "They tell me that if I shorten my two hours for lunch, I shall be able to make much money," a Frenchman is supposed to have said, "but what could I buy with all that money that would give me as much pleasure as taking two hours for lunch?" It was a question that seemed destined to stand insoluble for a long time.

The change in Paris was not just its Americanization. If the *grands boulevards* look too much like Broadway—Mr. Ybarra shook his head sadly—then wait until the visitor has a look at the Champs Elysées of 1927. From the Place de la Concorde to the Place de l'Etoile there was hardly a residence left. It was all just like Fifth Avenue, just a strip of surfaced roadway jammed with "speeding, honking, evil-smelling automobiles" running between a double border of candy stores, automobile showrooms, millinery shops and hotels. Ybarra blamed that change on the automobile and the "shameless dealer" who fifteen years before had exhibited a car in a window on the Champs. After that came the deluge.

As Paris began to choke on automobiles, an even newer mode of transport was making its beginnings. Lowell Thomas, who had made an airplane trip around the world in the mid-1920s, now went off to Europe, taking his wife on a new sort of voyage, a flying holiday. It was novel enough for Houghton Mifflin to publish Thomas's book on the subject, a 1927 tome entitled *European Skyways*. Thomas wrote, "It seemed to me if men could circumnavigate the globe without even changing airplanes and live to tell the tale, surely flying was now safe enough to invite one's family to join on a tour of the organized airways of Europe." Mr. and Mrs. Thomas took the boat to Europe and endured the chiding of their fellow passengers. "I'd rather swim the Channel than fly it." "Aren't you afraid you'll bump into an Alp?" "Sounds like a circus stunt." "Have you left arrangements for your funeral?"

Undaunted, the Thomases set out from London Terminal aerodrome for the four-hour flight to Amsterdam. Five European countries were already flying out of the airport: Britain's

Imperial Airways, French Air Union, Deutsche Lufthansa, KLM and Sabena.

The Thomases discovered that the red tape that was so irksome when crossing the border by train or car had not yet found its way into the aerodrome. Passengers were looked upon with respect and consideration, perhaps because they were considered intrepid pioneers, just back from or on their way to another perilous voyage. Travelers could take sixty pounds of baggage with them, but anything more weighty would go by freight plane. Airlines picked up trunks at the hotel and delivered them at the onward destination a few hours later. The service reached as far as Budapest and Warsaw.

Thomas's plane, which took them on the leg from London to Amsterdam, had room for fourteen passengers, seven on each side of a narrow aisle. The pilot and mechanic, in fur-lined suits and boots, leather helmets and fur-edged goggles, sat in an open cockpit up front. The cabins could get frightfully stuffy and the motor noises were deafening. "Two hours of the quivering throbbing vibration of those tremendous motors" wrote Thomas, "exercises and massages as many muscles as a ride on President Coolidge's electric hobby horse."

Although H. G. Wells, an early critic of early airline travel, had called the air service "unpunctual, untrustworthy and inconsiderate," Thomas thought planes were more inclined to run on time than trains. He and his wife logged 25,000 miles from London's Croydon Airport to Amsterdam, Paris, Budapest, Istanbul, Berlin, Moscow, Helsinki, Stockholm and back to Paris.

While Thomas's trip could not exactly be called stylish traveling, still commercial aviation had come a long way in five short years. A traveler named Ruth Osborne Ewan, who flew the Channel on a commercial plane, described it this way in a book she wrote in 1922:

> At Croydon we were dumped out of the automobile before a wooden building into which we plus our baggage were directed . . . with our 30 pounds of baggage each we stepped out of the opposite door. There they lay—

three great machines, ready for the trip. Excited mechanics rushed about tightening a bolt here, testing a wire
there, and oiling everything except us and our suitcases.
. . . We were scarcely in our comfortable wicker chairs
when the engines started. There we were out in the nose
of the machine feeling rather helpless and suddenly sure
the channel ship wouldn't have been such a bad way to
cross after all. There were just four of us in the little front
compartment. Just behind that sat the pilot and his mate
and back of them was another compartment for eight.
The channel trip took two hours and forty minutes of
flying, but door to door required five hours.

Europe had gotten into the airline business immediately
after the war. Oddly enough the Germans, newly surrendered
and soundly beaten, were first off the blocks. Less than a hundred days after the Armistice was signed the Deutsche Luft-
Reederei had opened commercial passenger service between
Berlin, Leipzig and Weimar. It took the British and the French
until the end of 1919 to begin scheduled service across the
Channel between London and Paris. Passengers flew in a converted bomber wrapped up in helmets and flying coats. KLM
started that year, fevered perhaps by the presence of Tony
Fokker, who had built German warplanes. European countries
recognized immediately that the small passenger revenues
could not sustain the cost of maintaining the new airplanes.
The converted bombers were too small to carry a payload and
the customers were too scarce. The French undertook to pay
part of each commercial pilot's salary. The British finally followed suit, forming Imperial Airways, a government subsidy,
in the spring of 1924. By 1926 it had carried twelve thousand
passengers without a fatality.

There were accidents, of course, but everyone seemed very
jolly about them. Reported Allan Campbell Orde, a British pilot,

One of the blades of my wooden propeller started to come
apart. The fabric coating, which covered the leading edge
of one of the blades, had worn through and the first thing
I knew there was a terrific bang and very heavy vibration.

I had to come down on the beach at Le Touquet. We were short of daylight but my two passengers were very nice. I explained what the trouble was and one of them said "I've got a penknife on me. I'll have a go at fixing this thing." So he helped me, and we hacked this loose piece of wood and fabric off the blade and got it started up again. I took off from the sandy beach and we got across to England while there was still daylight. In those days the passengers took part in the whole thing in very good spirit.

Recalls another pioneer in flying, British Air Line executive Dennis Handover,

On the London–Brussels–Cologne service we were flying largely DH-34's which were single-engine aircraft carrying eight passengers and a crew of two. We didn't have stewards or anything of that kind. There were no refreshments. I don't think we even had a lavatory. We had a great number of sick cans, however; flying around unpressurized at low altitudes was very sick making. You never got above 2,000 feet because you didn't dare lose sight of land. We had a very, very amateur type of radio telephone. It was one of those things that were subject to atmospherics. It was a question of shouting down the microphone and hoping. Very often you couldn't hear whether they had received your message or not. I remember it took me very nearly a week to get from London to Cologne on one occasion because we had forced landings all over the place. But the passengers didn't leave the plane. It was an adventure in those days. Not like today —if you're ten minutes late, there's a row. No, no. We stuck to the aeroplane. It was part of the fun. They knew, of course, that it was still pioneering. It was an event if you could come back and say, "I've just flown London to Cologne."

Those who weren't rushing down to the travel agencies to exchange the bundle they had made on the stock market for a ticket to Europe had invested in a Model T ($300 F.O.B. Detroit) and were off to see the U.S.A. The Model A was still a far cry from a T-Bird. For one thing, it came only in black and the floorboards—carpets came later—were put together from the box that had contained the battery. Still they offered a chance to take off on a Sunday excursion, to roll out and see relatives who had felt themselves pretty well insulated from family intruders by distance. And when the Model T refused to work one could always work on it, secure in the belief that one had at last stumbled upon a therapeutic hobby.

Like all new gadgets, the car became a necessary symbol that established status. It was therefore quite a bit more than merely a mode of conveyance. In *Babbitt*, Sinclair Lewis conveyed his concept of the status of the car:

> In the city of Zenith, in the barbarous Twentieth Century, a family's motor indicated its social rank as precisely as the graves of the peerage determined the rank of the English family. . . . There was no court to decide whether the second son of a Pierce Arrow limousine should go into dinner before the first son of a Buick roadster, but of their respective social importance there was no doubt; and where Babbitt as a boy had aspired to the presidency, his son, Ted, aspired to a Packard Twin Six and an established position in the motor gentry.

Determining social status by automobile make was a lot more difficult in this past era because there were so many other models. The Moon, Cleveland Six, Dort, Scripps-Booth, Jordan, Grant Six, Kenworthy, Roamer, Owen Magnetic, Auburn, Bay State, Chandler, Peerless, Jewett and Locomobile were trademarks of note on the roads in 1919 and 1920.

"On the roads" may be a questionable term because roads of this era were almost nonexistent. A 1921 touring book gives the following instructions on driving:

> A most important item is the towline—either hempen rope, chain or steel-wire cable . . . tire chains should always be carried; likewise some single chains or mud

hooks. There are also several automatic "pull out" devices which enable one to drive through places that might otherwise be impassable . . . where mountain roads, sandy stretches and muddy places are to be met with, or where the condition of the road depends on the weather. A shovel with collapsible handle and a good camp ax often repay a hundredfold the trouble of carrying them. To some a compass may appear superfluous but the seasoned tourist commends it.

In June of 1920, the Hudson River Day Line took cognizance of the car and knocked it. Its ad for its ships said, "No tire trouble this way—run your car on the steamship and you'll really enjoy the trip. No dust, no dirt, no bad roads to bother."

While the motorcar opened up the travel horizons of the American public, not every reaction to rising car sales was favorable. By 1920, religious and civic forces were calling cars "houses of prostitution on wheels." The joyriders drove down the road, strumming ukuleles and consuming by the pint liquids which, as one veteran of the age recalls, "their radiators would have rejected." One wary observer wrote, "Violation of the motor vehicle laws and the Eighteenth Amendment leads straight down hell's highway to the roadhouse. The auto has put the front parlor on wheels and moved it beyond the city . . . and all other limits . . . and I don't know which is the horn of doom . . . the klaxon or the saxophone."

The nation, put on wheels by Ford and the rest of Detroit, forced the roads to widen, lengthen and harden. As the network of byways increased, the day excursions grew into overnight journeys. Cities were often too far apart, and too difficult for a man who had taken to the open road. The automobile had given the traveler a new independence. He was at last free of schedules and itineraries. He no longer had to rush for trains and ships. He could leave when he wished and start when he wished. Once on the road, and reveling in his newfound freedom, he preferred to abide by no fixed itinerary. He wanted to eat when he was hungry and sleep when he was tired. A new type of lodgment, the overnight camp, appeared. They were clusters of six or seven cottages spread around a large community house. Each bungalow was furnished, in a style more Spartan than grand, with a bed, a chair and a washstand. But most were unfailingly clean and airy, and as the sign on the road so invitingly said, you could turn in and get a night's sleep for a dollar. In the morning the patron could usually be counted on to provide a pitcher of water, and up in New England, which was beginning to make a good business of catering to the summer motorist, the owner of one roadside camp sent his lady guests off on the road with an armful of flowers plucked from his garden.

Yellowstone Park thought its tenting arrangements were ideal, but it bowed to what it called the "bungalow hunters," and built a number of cabins. Although regularly camping was free at Yellowstone and a charge was levied for the use of the new shelters, the bungalows were always filled.

A novelty when it first appeared in the West in 1925, the bungalow camp, as it was often called, became a rage. In the Southwest the bungalow camps were often given a Spanish air, made of adobe brick or styled like pueblos. An early seer named Henry Burhans, in the employ of the Denver Tourist Bureau, predicted that one day there would be cabins of bungalow camps where the motorist could be assured of standardized rooms and rates. From the vantage point of late summer 1927 Mr. Burhans also foresaw that bungalow camps would eventually be equipped with laundry and valet service, a central garage capable of making repairs and a restaurant. "There is no doubt in my mind that much of the future of automobile camping is to be found in the development of the bungalow colony," quoth Mr. Burhans. The Chamber of Commerce noted that there were already two thousand first-class automobile tourist camps in the country, a new $20 million industry.

With cars and motels, came that other by-product, the trailer, or, as it is rather grandly known today, the motor home. Redden Trailers of Chicago offered the first manufactured commercial trailers in 1918, but most of the first trailers were homemade, utilizing car wheels and frames, which, by 1920, were inexpensively available from secondhand shops. Some of these early craftsmen built their trailers on all four wheels of a car skeleton, realizing a great deal of load space but encountering the tricky problem of backing and maneuvering. Others hacksawed the rear axle and wheels away from the chassis and came up with two-wheel trailers. In either case, a platform was built on top of the wheels, a load put in place and provisions made for tying it down and covering it with a tarpaulin during bad weather.

Early devotees of the trailer included Henry Ford, Thomas Edison, Harvey Firestone and John Burroughs, who undertook a series of camping caravans in the early years of the new decade. Henry Ford built a portable electric plant to light Edison's incandescent bulbs, which were strung in all tents at each evening's encampment. When a camp was pitched, the meditative Burroughs would frequently have his tent placed in a glade apart from the rest, where he could meander, in linen duster and with long white beard flowing, among the local plants and creatures which, as a naturalist, he loved. At the same time, Ford and Edison—if Edison wasn't reading in the front seat of one of the touring cars—would walk along the edge of the stream figuring how much electricity it could produce. When the group came upon examples of small industries, Firestone would speculate about how modern methods could improve their production. Wherever they went, these early campers-for-fun made a lasting impression on the people of the community and were sometimes woven into local folklore for a number of years thereafter.

With cars came buses, the vehicle that signaled the eventual demise of the trolley car. During the decade from 1916 to 1926 about 2500 miles of electric trolley tracks were abandoned in favor of bus service. Wildcat operators popped up everywhere and the industry spawned such lines as the White Swan, the Jack Rabbit, the Whippet, the Tiger and the Golden Eagle. In 1921, the first real intercity type of bus was manufactured. Several of these coaches were painted gray and looked so slim and trim that people began dubbing them "the Greyhounds." Some of these buses operated out of Muskegon, Michigan, and to spread the fame of the name, the local company adopted the slogan "Ride the Greyhounds." As this line merged with larger organizations, the distinctive name of Greyhound was chosen because of its advertising possibilities and the name became a national symbol for dependable transportation. By 1925, there were more than 6500 bus companies in the United States. They operated over 7800 different routes, but each company had an average of only two buses. Operations were inefficient and in most cases fares were high. Attrition and amalgamation were eventually to cure both faults of the industry.

Between the automobile and the bus, America was opened to the casual traveler. Before Detroit gave the country the car, vacationists chose the mountains or the shore, decided on a resort and then settled in for two weeks. A day's excursion, if one counted on a horse and buggy, or, worse yet, a bicycle, was rarely farther than five miles out of town. But now almost all the United States was available for exploration. By mid-winter 1925 more than half the vacationists en route to Florida came by car; in Southern California the figure was about 40 percent. In 1925 Florida had already registered two hundred tourist camps. The rush to See America First, a movement necessitated by the war years, gathered momentum that lasted deep into the 1920s and 1930s. The drumbeaters representing states and resorts vied with each other for the country's attention, offering the best of climate, the most fascinating historical sights and the greatest scenic landscapes. From California came the telling query popping up often in the magazines, "Icicles or roses . . . which do your children see at the windowpane?" In the summer the New England resort keepers invited the stifling city dwellers to head north and "Sleep under a blanket every night." The Upper Peninsula of Michigan became known as Hiawatha Land. (Minnesota had already staked out a claim to Minnehaha.) Virginia cited its battlefields with explanatory tablets. New England dusted off the lore of the Pilgrims. Washington's overnight presence was advertised from New York to Virginia. The traveler who came by car renewed the country's interest in the romance of its history, made it conscious of the beauty of its scenery.

California was exploding with new residents come to stay forever, and new tourists come to soak up the sun for a few weeks. Twelfth in population and behind New Jersey in 1910, California had jumped to eighth in 1920, and was beginning to crowd Michigan. Gold and oil had brought dramatic wealth, but the sunshine of Southern California seemed to be proving a new and more enduring attraction. The nation, free to travel, wanted to go where it was warm. It had discovered Florida earlier, but the real Miami boom was just gathering steam now and the competition with Southern California was hectic. Both had been opened by the Spaniards, both had been settled early but discovered late. Orange groves had blossomed in both states. And both had maladies that were looked upon with some trepidation by the new land buyers: California suffered from earthquakes and Florida had seasonal spells of hurricanes.

But the earthquakes didn't impede California's growth and the hurricanes were hardly enough to blow away the incredible Florida land explosion of the 1920s. Lots were bought in the morning and sold in the afternoon at four times the price. Some tracts mushroomed 1000 percent in a week. Caught in a fever, determined to make a killing or at least to carve out a *pied-à-terre* for comfortable retirement, the land buyers were hypnotized with tales of the exotic tropical life, of the waterways that eddied past the front door of one's property, of drooping palms and perpetual spring.

The real-estate salesmen centered about Miami came to the office all dressed alike in white suits and knickers. They developed their own patter, dismissed cockroaches somewhat quaintly as "palmetto bugs," took orders by telephone. Their ranks were joined by such an illustrious collaborator as William Jennings Bryan, just back from the Democratic National Convention of 1924 that had elected John Davis as candidate for president. Bryan didn't carry an order book, but he was enrolled nonetheless to sit by the side of a pool in Coral Gables and declaim in mellifluous tones to the people in the grandstand across the water on the glories of owning a piece of land in these salubrious climes.

Beset by the boom and the bootlegging, civic institutions bent and wobbled. In Hialeah, later to become famous for its racetrack, two slates of politicians claimed power so both hired separate police forces whose members rushed about arresting each other for impersonation of an officer. Despite the near-anarchy in some quarters, the local governments stood behind the local promoters no matter what. Miami saw fit to add a line to a folder it was circulating at the time, which warned, "Miami does not stand for pikers or knockers or pessimists. The wide awake citizens run them out of the city." When a knocker named W. O. McGeehan, the sports editor of the New York *Herald Tribune*, came to St. Petersburg to follow the ball clubs in winter practice and chanced to drop in a poking line or two about the land promoters, the reprisal was swift. An editorial appeared in the local newspaper gently remonstrating with McGeehan under the headline "Shut Your Damn Mouth."

The newspapers, like everyone else in south Florida, were riding the coattails of the boom. At one point so much real estate was being offered that the Miami *Herald* was carrying more advertising than any other paper in the nation. Its competitor, the Miami *Daily News*, once put out a 504-page issue in 32 sections. The land boom increased the size of Miami from 30,000 in 1920 to 85,000 in 1925. Building permits, by this time, had risen over the $100 million mark.

Across the water, the growth on the nine-mile-long sand pit many years later to be world-famous as Miami Beach, was less impressive. It had been little more than a swampy tangle of mangroves in 1912 when a nursery man named Henry Lum planted 38,000 coconuts, and another named John S. Collins from New Jersey had tried his hand at raising avocados. Collins, with the help of his son-in-law, Thomas J. Pancoast, began to build a wooden bridge across the bay. Helped by the Lummus brothers, Miami bankers, and encouraged by the interest of Carl Graham Fisher, the bridge was built and the town incorporated. J. N. Lummus became the first mayor, Pancoast was designated the head of the first chamber of commerce, and Carl Fisher became the prime promoter. Later a hotel was named for Pancoast and, of course, Miami Beach's famous Collins Avenue was named for his father-in-law, the avocado planter. Fisher piled in money and energy, but still the progress was slow, and even by 1930 Miami Beach had hardly attracted 7000 residents.

Up the coast, Palm Beach was off to a better start. After all, some of its first settlers had been entrenched there since the end of the Civil War, and then, too, it had gotten a tremendous boost from Henry M. Flagler, who brought his Florida East Coast Railway into the Palm Beaches before the turn of the century and had planted huge hotels on Lake Worth and later on the beach. But even the Palm Beaches, solid as they were, were not secure from the land boom and the boom beaters. At war's end Palm Beach had been enlivened by a sometime architect named Addison Mizner who had come to Florida at the age of forty-five to die. The climate proved such a tonic that Mizner was revitalized and he was soon wooing high society. He built homes for the Stotesburys, the Phippses and the Biddles, all royal names of the day. He treated them with high imperious manner and built their homes with only casual reference to cost, but there are those who to this day call him a genius and give him credit for some of the best houses in Palm Beach. Soon Addison was joined by his brother Wilson, a glib promoter down from Broadway. Together they moved into the town of Boca Raton, 27 miles from Palm Beach. Addison, known as the Aladdin of Architects, needed only the hyperbole of his brother to promote a project grander than anything yet undertaken in those giddy Florida days. They would build "the Golden City of the Gold Coast." It was to be a Spanish city inhabited by millionaires, sprinkled with golf courses and threaded with canals. The Mizner house itself would be built on an island in the middle of Lake Boca Raton. It would be equipped with a drawbridge that could be elevated upon the approach of a bore, a creditor, or perhaps, an outraged client.

Addison did succeed in drawing T. Coleman Du Pont into his plan. Du Pont had been a general, albeit of the Delaware Militia, and Wilson liked the promotional possibilities of having a general in the company. Then, too, Du Pont had been president of the E. I. Du Pont de Nemours & Co., a very respectable personality for a letterhead. Wilson used the general's name in a newspaper ad, which sent land sales rising over $2 million a week and the general's blood pressure bubbling over the boiling point. When Wilson sent out a release that talked of guests disembarking from their yachts in front of the Ritz Carlton, the Mizners' newest hotel, the general thought he had gone too far. After all, the hotel hadn't yet been built.

Du Pont rallied the stockholders to eject the flamboyant brothers. For evidence he produced a police record purporting that Wilson had once been convicted of running a gambling establishment. No slouch at infighting, Wilson produced a pregnant and unwed mother-to-be who obliged by swearing she was carrying the general's child.

A magazine once called Mizner's idea of Boca Raton "a happy combination of Venice and heaven, Florence and Toledo with a little Greco-Roman glory and grandeur thrown in." The world's largest hotel, which they had planned to build on the beachfront, alas, never got built. Before it all went up in gilded smoke dreams, Mizner did manage to raise a modest gem called the Cloister Inn, a 150-room hotel which he filled with ceramic jars, beautiful tiles and curlicued furniture, all brought from Spain without thought to expense. He lavished $1.5 million, a sizable sum for the day, on the Cloister. He strung iron railings around the balconies and hung them over date palms and orange trees. He built six-foot-high walk-in fireplaces, tucked tiny little arched windows up among the eaves. When it was all done the Ritz Carlton Company was brought in to manage it. High society arrived, but Florida was coming on low times. In eight months the Cloister Inn went broke. Business wavered and shook, and as the final *coup de grâce*, the devastating hurricane of September 17, 1926, swept over Florida, blowing off rooftops and exposing all those new homes to a devastating rain. A hundred feet of the new Million Pier was yanked off and dropped into the sea. Three hundred died and a thousand more were injured. The Florida propagandists—of the same school that called cockroaches palmetto bugs—sent out the word that the damage had been slight. But the food and medical shipments that finally had to be sent from the North were proof enough. The great gasbag that was the Florida land boom strained and burst, and evaporated into the blue stratosphere.

For Miami it was the end of a phase, a time when the moguls were in royal residence: Harvey Firestone in the oceanside estate on the beach; James Cash Penney in his winter home; Gar Wood, a face familiar to the newsreels, with his burnished face and shock of white hair. There were Gene

Tunney training for the big fight, President Harding relaxing at golf, Paul Whiteman leading the orchestra, and the rum-runners racing the Coast Guard boats up and down Biscayne Bay.

Those days evaporated and so did the Mizners. Addison's great dream of the Golden City on the Gold Coast was purchased for ten cents on the dollar by Clarence Henry Geist, once a railroad conductor on the Blue Island out of Chicago. Befriended by Rufus and Charles Dawes (Charles was vice-president under Coolidge, later ambassador to the Court of St. James's), Geist, with the backing of Dawes's Central Trust Company of Illinois, became a public-utilities operator, eventually the largest one in the United States. Living in sunny comfort in a villa on Worth Avenue across from the Franklin Huttons, Geist came home one day and casually announced he had bought Boca Raton from seaside to silverware. The family moved up the coast, where Geist went to work transforming the Mizner idea into one of the fanciest private clubs ever built in the United States. He brought down Schultz and Weaver from New York, later to build the Waldorf-Astoria, had them put up 450 rooms. In the Cathedral Dining Room thrusting columns were coated with real gold leaf, topped with carved medallions that in turn upheld arches that disappeared into the vaulted ceiling. For the uninitiated coming into breakfast for the first time it all seemed a bit like ordering ham and eggs in a splendiferous church.

Tourism within the United States was, of course, not limited solely to the two playgrounds of California and Florida. During 1926, 2.25 million people visited the national parks, a gain of 36 percent over the total of 1924. Yosemite was the prime attraction, having 275,000 visitors. After that came Hot Springs, Arkansas, Rocky Mountain Park in Colorado, Yellowstone, Mount Rainier and the Grand Canyon.

By now America had developed a talent for merchandising. Already adept at selling soap and cars, the entrepreneurs turned, with grand Barnum flair, to selling dreams. Resorts, as well as ocean liners and trains, vied with each other for the traveler's patronage. One of the greatest stunts ever conceived was Atlantic City's Miss America Beauty Pageant. Its origins were mild enough. At a meeting of newspaper circulation managers in the winter of 1921, Harry Finley, representing an Atlantic City paper, proposed an idea that would help build readership, a matter of prime importance to those attending. Papers in cities in the East would run a popularity contest among young women. The prize for the winner—since Finley had thought up the scheme—would be a trip to Atlantic City.

A program of events was concocted, among them a "National Beauty Tournament." It was of little matter that the contestants hardly represented a nationwide array. The entrants were to parade in bathing costume which, in the fashion of the times, included long black stockings. A local reporter, Herb Test, offered the ultimate hype. The winner would be called "Miss America."

By 1922 the Atlantic City Pageant had become an extravaganza with baby parades, a display of flower-decorated rolling chairs, and a "Bather's Revue" with everyone, including the band, attired in bathing suits. The grand finale was the bathing beauty contest to pick Miss America. By 1923 there were fifty-seven contestants, all selected by newspapers. The blizzard of clippings put Atlantic City on the map as nothing else since salt water taffy became the resort's trademark confection.

With all that notoriety Atlantic City assumed a new popularity. Some easterners came in winter and wrapped themselves in steamer rugs and took the pale sun (and the chill Atlantic winds) supine in a deck chair. It was almost like going to sea, and besides, it was handy to Philadelphia and New York. Others went to Florida where the indomitable Henry Flagler had built the Overseas Railroad, stretching over the one hundred miles of keys and water to Key West, which liked to call itself then "the only frost-free city in the United States." The rail journey from New York to Key West via the Overseas Railroad took forty-six hours. The intrepid could continue onward to Havana aboard the Peninsular and Occidental Steamship Company's ships sailing daily across the ninety miles to the Cuban capital. Key West was also connected by boat to Tampa during the winter, and the Mallory Line, plying between New York and Galveston, called at Key West en route.

Flagler's dream had included an elegant hotel at Key West, and in 1921, eight years after he died, the Casa Marina, latest of the Florida East Coast Hotel Company's establishments, opened with flapper-era fanfare on Key West's southern shore. Guests were driven to the porte cochere and ushered into the beamed-ceiling lobby set with rattan chairs. They danced in the ballroom and took their meals in the dining room which, in the traditional style, was a mass of tables lined up in military formation. Guests could angle for "salt water fish," as a booklet of the times called deep-sea fishing, play tennis on the hotel courts, or, beknickered, show up for golf on the nine-hole course the city had built. "The weather is such," said the Casa Marina's brochure, "that baths may be taken during any day of the entire year." One had only to cross the lawn to find the sea.

The Cuban extension was the first significant foray into the Caribbean. Other travelers boarded United Fruit Company ships, which dropped their passengers at Port Antonio in Jamaica. While the ships took on bananas, visitors would enjoy a vacation at the company-built Titchfield Hotel. Later they would be taken to Kingston and the Myrtle Bank Hotel to reboard. Many company officials also used the hotels for vacations in Jamaica. By 1924, six steamship companies were supplying regular services to Jamaica and bringing winter visitors.

In the midst of this national preoccupation in pursuit of the good life, a handful of aerial swashbucklers—mostly ex-Army pilots turned barnstormers—were paving the way for a booming commercial aviation industry. The First World War had left many young men interested in flying, a topsy-turvy business at best. During the 1920s, the greatest hazard in flying was often described as the risk of starving to death. Gradually, however, a new science was growing up. The globe was circled by air in 1924 by the U.S. Army, a feat that took 175 days. The same year the Army sent a plane across the country, flying from coast to coast in 21 hours and 48 minutes.

Although Europe was off and flying almost as soon as the guns were laid down, the art of carrying passengers by air for money had a more ponderous beginning in the United States. American had to cope with the problems of long distances and the stout objections to federal subsidy. It really wasn't until the late 1920s until city to city service started. Vacationers were the first benefactors of the early air routes in the United States. As early as the summer of 1919 a company called Aero Ltd. began flights between New York and Atlantic City. Summer commuters and pleasure travelers were the principal customers, but the pickings looked better in the South. Aero Ltd. moved to Miami and began Prohibition airlifts between Miami and the oases in the Bahamas. Aeromarine Sightseeing Navigation started a similar service between Miami and Havana.

The fear that federal subsidies would suffocate public initiative and lead to more governmental controls was finally overcome by a simple device. Washington agreed to give airmail contracts to private firms. It was an airmail pilot, Charles Lindbergh, who finally got U.S. aviation going. Lindbergh's solo flight to Paris in 1927 not only won him the $25,000 Orteig prize, but it made him the hero of the nation. He got 100,000 telegrams, 3.5 million letters containing over $100,000 in return postage. He had his pick of over 7000 jobs and an offer of a straight $1 million if he would get married in front of the movie cameras.

Lindbergh disdained all the jobs and offers of marriage, but he did join Transcontinental Air Transport and helped to map routes, set up airport standards and create schedules. TAT became known as "the Lindbergh Line," the forerunner of TWA. It sent its first trimotors across the United States in July of 1929. The Ford airplanes could carry ten passengers at a speed of 110 miles an hour, enough in those times to make transportation history. Now one could cross the country in a combination of air and rail trips in 48 hours at a cost of $351.94. Inspired by Lindbergh's daring flight into Le Bourget, American aviation now began to attract both backers and customers. Not even the oncoming depression could slow its flight into the 1930s.

The decade that had begun with the gay abandon of the 1920s ended with the ominous clash of cymbals that announced the collapse of Wall Street in 1929. The roof had fallen in and the Depression was on. Down in Boca Raton Clarence Geist's posh club for swells lent money to any guest who had had no trouble a few months before in meeting the $5000 initiation fee, but now needed carfare home. Many of those guests who fled back north were never seen in the Southlands again, for the financial ruin was too much to bear and they flung themselves from high windows into the canyons of the cities.

It had been a delirious decade, the whole ten years an almost dissolute frivolity. It had spun too fast, flown too high, and now it was over, too soon. Short as it was, it had been a time when Americans, seasoned by a foreign war, inspired by their new global interests, supported by their new wealth, strode out across the world to see how the rest of it lived.

> Travel has changed from a sport to an industry [wrote Mildred Adams in the New York *Times* two years before the crash], with travelers as its customers, all known means of transportation and housing as its machinery, and scenery, romance, and adventure as its products. It reaches into every hemisphere and every country. It can quote hotel rates from Tierra del Fuego to Baffin Bay. It knows the rate of exchange on shekels and kopecks and the fastest train accommodations between Cape Town and Vladivostok. Its richest rewards go to the men who can discover new combinations, new lures or new ways of satisfying old desires.

The new lures and the new ways would have to wait while some elements of the once affluent society set up apple stands on the street corners and others stood in line to cadge a bit of free warmth from a soup kitchen. But the fire had been lighted and the flame—though it might flicker and falter in the desolate times ahead, would never go out. Mildred Adams wrote:

A thousand threads of desire catch at Milwaukee flour men and Nevada miners, at New York movie magnates and Dartmouth students. Places read about are so alluring that many a Mid-West spinster has cherished dreams of Dickens' England or Rider Haggard's Africa for an entire lifetime of gentle drudgery. . . . History, romance, religion, art—all these move men's souls to longing and their feet to traveling. . . . There was a boy in Missouri who bought a small piece of strange carpet because his eyes could not escape the lure of its soft colors. He asked merchants and hunted in libraries until he learned the place it came from. When at last he sailed away it was for none of the familiar ports. He was bound for a small seacoast town from which he would go by camel to the tiny Syrian village where his carpet was woven.

When they had money again the citizens would be sailing again, for travel, travel agents and the travel industry had all become a big business.

VI | *Can You Spare a Dime? (1930–1939)*

"POVERTY WILL BE BANISHED from the nation," assured President Hoover that March day in 1929 when he took office. "The future of the country . . . is bright with hope." That summer Bernard Baruch told a friend, "The economic condition of the world seems on the verge of a great forward movement."

And yet there it was October. On the twenty-fourth of that memorable month 13 million shares of stock were sold, and on the twenty-ninth the bottom fell out. "When Wall St. took that tailspin you had to stand in line to get a window to jump out of . . . ," wrote Will Rogers in a comment that wasn't funny to everybody. It would not be long until the bread lines formed in Times Square. Soon there would be 16 million unemployed in the United States. Men without the price of a want ad and perhaps without the price of a newspaper strode the streets wearing placards that read, "Job Wanted."

It was the worst of times for a big expensive luxury hotel, but there was President Hoover in 1931 sending out an announcement of admiration over the radio for the brave group who had built the new Waldorf-Astoria.

It was the last day of September 1931, and those who had received special invitations crowded around the ballroom entrance waiting for the Girl Scouts to show them to their seats. Oscar of the Waldorf stood at the head of the Park Avenue stairs just as he had stood at the head of the stairs of the old Waldorf when it had opened on the site now occupied by the Empire State Building nearly forty years earlier. A quiet fell

162

over the crowd as the President's voice, beamed by NBC from the Cabinet Room of the White House, came over the loud-speakers. "The erection of this great structure has been a contribution to the maintenance of employment, and an exhibition of courage and confidence to the whole nation," he said. He recalled that when the City Hotel had opened seventy-three years before it had been called the first of the great hotels and people had come to look at it and marvel at it as a symbol of the nation's growth. The City Hotel had fewer than a hundred rooms and now here was the Waldorf, forty-seven stories high, with 2000 rooms, 1600 people on the payroll and a private railroad siding in the basement.

It was altogether fitting that across town at the National Theater, *Grand Hotel* was the hit stage play. *George White's Scandals* featured Ethel Merman, Rudy Vallee, Ray Bolger and Willie and Eugene Howard, and Ginger Rogers was on the stage of the new Broadway Theater at Fifty-third Street. On October 1, 1931, 500 guests awoke to the Waldorf's first morning. It was the best house count that it was to ring up for some time to come.

The Crash riddled the ranks of Palm Beach, which had been for three decades a nesting place for bankers and captains of industry. Up at Boca Raton, Geist's guests had gone home, paying their way with the loans which the club provided from the $25,000 in cash it happened to have on hand. And what of Geist? There he was with one of the most expensive, one of the most exclusive clubs in the world at the beginning of one of the worst depressions the world had ever known. He had called in Schultze and Weaver, who had designed the new Waldorf-Astoria, ordered 450 bedrooms and baths to be built, planted 450,000 trees.

Somehow Geist weathered it all and kept the Boca Raton Club. Its highly exclusive membership never faltered. Geist ran it all like a martinet, a martinet complicated by complexes. He was a virulent anti-Semitic, having worked up a hatred that was nearly pathological. Any member committing the indiscretion of bringing a Jewish friend to lunch was expelled at the door and handed back his membership fee. Geist ordained that the club's moving-picture show should not begin until he and his party were seated. Plagued by debilities and living in an age that had not yet produced the golf cart, he had roads built around the golf course, had himself driven over the links in his chauffeured Rolls-Royce, a rolling clubhouse equipped with two sets of clubs, four sets of shoes and a bottle of Old Grand-Dad. He died in 1938 having amassed in his lifetime what would be, if measured in comparative purchasing power today, an estimated half billion dollars. He spent $8 million of his own money on Boca Raton, and left the club $100,000 a year to keep the doors open.

A private club with an angel the likes of Geist might survive, but lesser hotels, bolstered by no such slush fund, even unimpinged by Boca Raton's selectivity clauses, were having a hard time seeing the era through. Along New York's Broadway, 70 percent of the hotels were broke and empty. Guests had checked out owing bills that totalled over half a million dollars. Most managements began to protect themselves by issuing bills every three days. Soon "due bills" appeared, a certificate tendered in lieu of cash, bartered for anything from groceries for the hotel kitchen to newspaper and magazine advertisements that might coax the customers. Few guests paid cash at hotels, not when they could get a due bill for a 40 or 50 percent reduction. Some agencies dealt exclusively in due bills, buying them from purveyors who had received them in barter, and selling them to commercial travelers and those looking for a cut-rate vacation.

In Atlantic City the marble banks became marble ginmills. The day after F.D.R. closed the banks in 1933 businessmen and even bankers, spending winter vacation in Miami Beach, found themselves without ready cash. Nor could they write checks on their frozen accounts. There was nothing much to do but make a grand exit through the front door without paying their bills. After that interim crisis had eased, the faithful came back to Miami Beach again, but most of them had money only for short stays. They carried home stacks of embossed hotel stationery, wrote their letters in the frozen North, and sent them south again to be mailed from Miami Beach. You had to keep up a good front.

If conditions were bad in America, they were miserable in

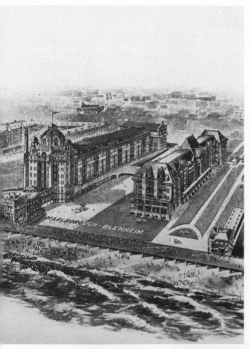

When the railroad arrived in Atlantic City in 1850 there were only seven houses in town. But the boardwalk, laid out twenty years later, was soon called "the most fascinating boulevard in the world." By the late 1880s, the seaside resort had become the great democratic resort of the Jersey shore where "children of Philadelphia millionaires and children of Absecon oystermen ride together on the merry-go-rounds," as one newspaper put it. The grand double hotel, the Marlborough-Blenheim, is shown as it appeared about 1912. The view of the boardwalk, top right, with its famed wicker strollers, was photographed about the same time. The salt water taffy legend was born when heavy Atlantic rollers swamped a candy store, prompting its enterprising owner to advertise his sea-washed confections as "salt water taffy." Bathing beauties, above right, were 1921 Miss America entrants.

By the time the century turned, Americans were on the road journeying in all directions. "See America First" was a popular theme, and the first daredevil embarked on a cross-country tour in 1903. It took him 52 days to reach the West Coast. The stagecoach, upper left, carries visitors in Yosemite; the curious excursion car toured Mt. Lowe, California. Those who went clear to China reported rickshaws more comfortable than those in Japan.

Hawaii's first hotels were in downtown Honolulu, not on the beach. The roof garden of the Alexander Young Hotel was a popular place where one appeared—not in splashy shirts and muumuus—but in staid tropical street attire. Lei sellers lined the sidewalks, and a Waikiki surfer of 1902, coolly clad, found himself all but alone.

As the dauntless soared aloft—the new way to go—baggage stickers such as this one from Western Air Express became a mark of distinction. This line used six kinds of planes.

Americans were tempted to travel in assorted directions. This advertisement, below, lured snowbound easterners to California, while the saucy girl in the boater made a fetching cover for Holland-America's passenger list.

Catalina was a handy romantic isle for Californians as this cover for the sheet music of a song clearly indicated. Band music became a radio standby.

THE PACIFIC MONTHLY

HOLLAND-AMERICA LINE

THE GREENBRIER GREAT NORTHERN RAILWAY

In 1928, 437,000 Americans sailed abroad. But many foreign notables came here, among them a soulful Prince of Wales done up in plus fours for a golf game at the Greenbrier, above left. A feathered party from Rumania visits western Indians: Queen Marie is flanked by Princess Ileana and Prince Nicholas. Honolulu sailing day, below, was a gay rite.

ARCHIVES OF HAWAII

Motoring was off to a rude start in America, where the early drivers were called "autoneers." Constables disguised themselves as workmen and strung rope across the road to trap any autoneer doing more than eight miles an hour. Early tent trailer, top, was a forerunner of the motor home. Model T "truck house," above left, appeared in 1924; de luxe trailer, above right, in 1929. Cars, lower left, were parked in Yosemite on Memorial Day, 1927. Hertz rented cars in Florida as early as 1925.

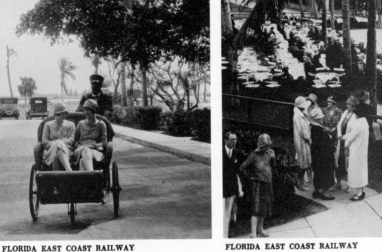

FLORIDA EAST COAST RAILWAY   FLORIDA EAST COAST RAILWAY

The sun lands of both coasts were attracting visitors escaping the winters of the Twenties. The Mizners were in Florida building houses and hotels, and these cloche-wearing ladies were en route to the Everglades Club aboard what was called an Afromobile since it was propelled by a driver of African heritage. Afternoon socials were in vogue at the famed Coconut Grove of the Royal Poinciana in Palm Beach. Out in Venice, California, meanwhile, these sightseers rode an electric sidewalk car.

That indefatigable railroader, Henry Flagler, stretched his tracks over 100 miles of keys and water to create the Overseas Railroad, top left, bringing passengers to Key West. Before the crash Europe was a playland for Americans. A leisurely sea voyage was quite in vogue. So was traveling abroad for business. The Swedish American liners, *Kungsholm* and *Gripsholm*, upper right, sailed to Gothenburg from New York. By 1929 the noise of the airplane was heard. Ford Trimotor, below, seated passengers in wicker chairs.

Germany. Communism seemed attractive, but then so did Nazism, the counterforce which was supported by foreign and domestic money eager to hold the communist wave in check. Said Christopher Isherwood of Berlin in the waning season of 1931, "Like a long train which stops at every dingy little station, the winter dragged slowly past. . . . Berlin was in a state of civil war. . . ." Hitler abhorred modern art, modern music, and Jews, Masons and Catholics. His adherents dutifully defiled the theater where Kurt Weill's operas were playing, forced art galleries to close and set upon the Jews. The year the Waldorf opened in such splendor in New York, 15,000 fell wounded on the streets of Germany.

It was a time to stay home and a time to think small. The Austin appeared, a pint-sized car, that reminded the Americans, and tickled them with a sort of irony, of how the prosperity of the 1920s had shrunk. And then there were miniature golf courses where you could play a game outdoors, hold a golf stick in your hands without bothering to join a club, pay a stiff greens fee or tip a caddy. It had all started when a man called Garnet Carter built a Tom Thumb Course on Lookout Mountain in Tennessee. When that was successful he built a second in Miami and the craze was on. A ride in the family car and a pause en route to tap a golf ball along a felt greensward was almost as good as a weekend in the country, and it passed for one in the 1930s.

While it was embracing small cars and small golf links, the American public also greeted the motel, a sort of shrunken roadside edition of a hotel, shorn clean of such embellishments as bellhops, grand lobbies and registration clerks in carnations and cutaways. An outgrowth of the tourist camps of the 1920s, they were run by Ma and Pa and perhaps a country girl to make the beds and clean the rooms in the morning. Anybody could check into a motel with anybody else, and frequently did. Frank Capra used the motel in his award-winning film *It Happened One Night*, which starred Clark Gable, Claudette Colbert and motels. In one memorable scene the two stars, still unmarried, fell asleep on separate cots divided by a bedsheet hung as a partition. They called it the Walls of Jericho, which, when love finally triumphed, naturally came tumbling down.

Off touring on the open road, Americans discovered their own country. They discovered the desert and they discovered the mountains. Hollywood finally happened upon Palm Springs which, after all, had built a hotel of its own as early as 1886 when Dr. Welwood Murray opened a 26-room inn. Dr. Murray bought his property from Judge John McCallum, the town's first white settler, who had come two years earlier looking for a better climate for his tubercular son. By the time the 1930s had arrived the Desert Inn, a famed establishment later to be owned by Marion Davies, had already been in business for better than twenty years. In 1932 a young movie star named Charles Farrell bought a few dusty acres out of town on a wide spot along Route 111. Two years later he thought it might be fun to build a tennis court. Soon the movie people were roaring out into the desert in their Dusenbergs, a four-hour-and-then-some excursion that ran a gamut of 226 traffic lights. With Ralph Bellamy, Farrell formed the Racquet Club, and when they put in a swimming pool the next year the retreat in the cactus became so popular they were soon enlarging the dining room.

Small resorts also began to spring up in the desert to the east, especially in the suburbs of Phoenix. Then came the Arizona Biltmore, exuding the airs of Frank Lloyd Wright, who maintained a winter workshop nearby at Taliesin West. The Arizona Inn, all by itself in the lonely desert when it was first built, would later have to surround itself with a high wall encircling its twelve acres and defending it from the encroachments of Tucson, which has sprung up around it.

Hollywood was at last developing its own oases within driving distance of Los Angeles, but wateringholes had long since proliferated along the eastern seaboard. None had grown more popular than the Borscht Circuit of the Catskills. Excursions to the cool air of the mountains had been popular since the days the steam trains were met by horse-drawn coaches. By the late 1920s and early 1930s the hotels and boarding-houses had become highly organized resorts that featured great gobs of chopped liver, borscht and sour cream and a dazzling program of entertainment. Many members of the social staffs who staged the shows, ran the games and kept the guests in a perpetual state of high glee were to become the most famous names in show business. The Flagler Hotel, locked in mortal combat for customers with Grossinger's, built a little theater complete with electrical switchboard, professional stage machinery and seats for 1500. Its social staff of twenty-six was headed by Moss Hart, in a few years to emerge as one of America's most celebrated playwrights and directors. His first assistant was Dore Schary, and his arch rival, the head of entertainment at Grossinger's, was Don Hartman. Schary became head of Metro-Goldwyn-Mayer, and Hartman head of Paramount Pictures. In the years that were to follow, the Borscht Belt and its southern branch, the Poconos in Pennsylvania, produced an incredible list of great American performers, among them Danny Kaye, Jerry Lewis, Eddie Fisher and dozens of others.

Some of America managed to get out of doors, too, embracing the new passion for skiing with great zeal. Man had been skiing for at least 4500 years—or so the archeologists in Sweden purported to prove—but it was not until the Nordic events were added to the winter Olympics of 1924 and the Alpine events to the Games in 1936 that the fever began to spread. Ski lodges sprouted in the middle 1930s, especially in New England, New York, Pennsylvania, in Quebec's Laurentians, and in California and Oregon. Some industrialists with a modest amount of money still on hand invested in rope tows and lifts. Ski trains and snow buses left on weekends to bring the snowbirds to the snow. St. Moritz, Davos, and Chamonix had been catering to winter visitors for years, but not until Averell Harriman, then chairman of the board of the Union Pacific, decided to build Sun Valley was America to have a true winter sports resort of its own. Sun Valley was set up in a snow bowl not far off the route of the Union Pacific. Harriman brought in Steve Hannagan, the Babe Ruth of ballyhoo, to make it famous and he did. Hannagan graced Sun Valley with movie stars and famed writers, made it twinkle with celebrated names. Pictures of its guests bobbing in the steam of its outdoor swimming pool in the middle of a snowbound winter made papers all over the country. So did scenes of Gary Cooper, who came to hunt, Ernest Hemingway, who came to write, and Sonja Henie and Glenn Miller, who came to do a picture called *Sun Valley Serenade.*

Sometimes the celebrities tripped over the goggle-eyed guests. They still tell the story out in Sun Valley of the stranger who asked Hemingway for his autograph. "Thanks, Mr. Hemingway," the pleased visitor said cheerily when he got the signature. He passed the table again and said, "Hullo, Mr. Hemingway." Intoxicated by being so near the great, the stranger permitted himself one more walk-past, this time calling out, "Hi, Ernest!" Further giddied by Hemingway's acknowledging nod, the man made one more pass, this time calling out, "Hello, Papa!" Hemingway lowered his beard and raised his arms. "Hellooo!" he roared, "and good-byyyyye!"

Hemingway was later to shoot himself at his home in Ketchum, and he is buried in this old lead-mining and sheep-herding town. Here in its brassy early days came the Hollywood stars and the elegant sporty types, looking for an evening's entertainment at the Christiana Gambling Casino or the crap tables in the saloons along Main Street. "They'd come down on crutches, some of them, and stand around gambling for hours," confided the boniface of the Pioneer Saloon to a correspondent of the New York *Times.* "The only heat they'd have in these places was a potbellied stove. Steak was eighty-five cents. Oh, my, it was gay!"

Some of America's traditional resorts tried to meet the new times. Greenbrier, in West Virginia, which had been a favorite resting place of Robert E. Lee, turned its racetrack into an airport in 1930. The next year, despite the nationwide shortage of cash and paying guests, it added three new wings to the main building. Up on Mackinac Island, the Grand Hotel still held out firmly against the automobile, insisted its guests walk, pedal a bike, ride a horse or engage a horse-drawn carriage.

In big resort hotels or big city hotels the air in the dining salons was subdued. There were no saloons nor would there be until Prohibition was repealed deep in the winter of 1933. Hit hardest by the drought, the city hotels had tried to recoup lost bar income with fancy soda fountains. They built satin-lined tea salons, hired gypsy orchestras, but it was no good. At the old Ritz, Albert Keller, one of the most celebrated maîtres d'hôtel of the day, had long since closed the famous Crystal Room. "Our patrons did not want a dry supper and a dry dance afterwards. . . . [It] was altogether too dry a time. A change has come over Americans since Prohibition. Every public assemblage seems to be merely a place where people go to be bored."

Egged on by such aficionados as John Mason Brown and Lucius Beebe, the country took once more to the rails. The railroads were getting better, inspiring these men of taste to erupt into ecstasies of rapture when they heard the train blow. "I like trains so much," wrote John Mason Brown for *Vogue*, "that were Mr. Rockefeller to remember me in his will I would be tempted to exchange my bed in my apartment for a berth, have it draped with heavy green curtains, equip it with a machine to rock me gently while I slept, and insure slumber by installing a soundtrack to release such blessed night sounds as distant whistles, expiring radiators, clanging road signals, passing freight cars, and the hum of ties. . . . I would have cinders blown in my face just to remind me of the good old days. . . ."

Beebe could lapse into sweet odes about the parade of westbound trains that whistled over the New York Central's tracks each afternoon at five on their way west. He pulled all the stops for the *Twentieth Century Limited*, the all-Pullman, extra-fare flyer which carried barber shop, valet and maid service, porterhouse steaks, flowers on the table, and a conductor with a pink carnation in his buttonhole. Of the *Twentieth Century*, the train that was to inspire a stage play (and later a musical comedy), Beebe was later to write:

It was a train apart, aloof, serene, incomparable—the sum of all excellences wrapped up in all-Pullman, extra-fare schedules that none might let or hinder. It was at once a force of nature and a national show piece. When it sailed each afternoon from its terminals in New York and Chicago and the red carpet of its going was rolled up for the night, each of its several sections became a self-contained microcosm of security, composure and the best of everything.

People of consequence rode it as a matter of course. People of diminished importance rode it to be associated with people of authentic stature in their various worlds. No guest list or passenger list anywhere—neither the register of The Palace Hotel in San Francisco nor that of the old *Mauretania* when she was the blue-ribbon vessel of the Atlantic sea lanes—boasted more effulgent celebrities. Women of fashion who lunched one day at the Pump Room and the next at The Colony dined on The Century in between. Tycoons en route from a director's meeting in the Wrigley Building to a finance committee at the Downtown Club read *The Wall Street Journal* in the club car of The Century. On days when its arrival in Manhattan coincided with the sailing of the Oylmpic, Mauretania or Berengaria, The Century was awash with international travelers with Vuitton luggage and the labels of the Crillon and Savoy Hotels on the hatboxes.

The nation was returning to the rails with such enthusiasm that American Express organized the first of its Banner Tours of the West. Special trains carried the prepared tours over a seven-thousand-mile itinerary at a rate of about two and a half cents a mile. Beginning with four tour trains a summer in mid-decade, the agency was soon dispatching a train a week, and then, before the war ended it all, had increased it again to a total of twenty-two excursions every summer.

While the railroads were laying on lounge cars and observation cars, introducing color by such designers as Raymond Lowey and Henry Dreyfuss, all-room trains, telephones between cars, speedometers and parlor furniture, anyone who took the airlines in the early 1930s had to have the disposition and durability of a pioneer. Timetables had little more relation to fact than twice-told tales. Speeds rarely edged over a hundred miles an hour. The low altitudes were bumpy and the ventilation systems were poor. Many airfields were muddy cow pastures miles from town. Business was so slack that one airline ordered its office workers to take two flying round trips every week to convince airport sightseers that the airlines were indeed attracting a lot of traffic.

The airport in Newark, New Jersey, serving New York City proved to be busier in 1933 than Tempelhof, Le Bourget and Croydon combined, but storms, navigational errors and mechanical failures made black crash headlines around the country. The nation remained wary. When Will Rogers was killed in a plane piloted by Wiley Post, one of the best pilots in the country, the risks involved in air travel seemed hardly worth the thrill of speed. *Hell's Angels*, which starred Jean Harlow and Clark Gable, only helped convince the public that flying was a sport for dashing heroes inspired by blond bomb-shells, not the average man.

In the background, however, the nation's aviation designers and engineers were working hard and long to develop the new techniques and devices which would make flying a swifter and safer form of transport. Trans-Western Airlines in 1934 intro-duced the Sperry Automatic Pilot—"Iron Mike"—on its planes. This tiny black box of intricate controls allowed the pilot to take his hands off the wheel and could actually fly a plane in level flight for hours at a time, reducing pilot fatigue and making longer hops possible.

On the ground, radio beacons were designed and improved. By listening for the beacon's radio signal in his earphones a pilot could tell whether he was flying to the right or left of his predetermined course. These radio beacons were literally high-ways in the sky which the pilots could follow from one city to another through clouds and storms.

Airfield runways were lengthened, paved and lit to make landings easier. Airport control towers were built. Traffic tech-nicians speaking over the radios learned how to clear planes for landings and takeoffs, minimizing the risk of crash near the landing fields.

In 1932 Douglas Aircraft was asked to build a metal plane that would be powered with three engines, carry a crew of two and a dozen passengers on a leap of 1000 miles at a speed of 150 miles an hour. A year later Douglas came up with the DC-1 (for Douglas Commercial). It didn't have three engines like the Ford monoplanes, it had two. It was all metal, a little larger and a little faster than Boeing's 247. It had flaps to slow it for low-speed approaches. The tests were so successful that

TWA, which had first suggested the plane, ordered twenty-five. With a few modifications Douglas began producing the DC-2. They were sold at $65,000 each, a price that lost money for Douglas but launched the era of the modern airliner.

Douglas's success with the DC-2 attracted a new client, American Airways, which had been using the Curtis Condor in its transcontinental sleeper service. It wanted a faster version for a New York–California sleeper that would fly through the night and arrive the following day. Douglas evolved the first of the DC-3s. The version it produced for American allowed room for twenty-one sit-up passengers or fourteen in the berths of the "sky sleeper," as it was called. Before the decade was out DC-3s were carrying 75 percent of the nation's traffic.

The Tumultuous Thirties were hardly two months old when S. A. Simpson, a San Francisco-based agent of the Boeing Air Transport Co., sat down and wrote a memo to W. A. Patterson, then head of the Boeing Air Transport Co., in Seattle.

As a suggestion—I was just wondering if you had ever given any serious thought to the subject of young women as couriers. It strikes me that there would be a great psychological punch to having young women stewardesses or couriers or whatever you want to call them, and I'm certain that there are some mighty good ones available. I have in mind a couple of graduate nurses who would make exceptional stewardesses.

Imagine the psychology of having young women as regular members of the crew. Imagine the tremendous effect it would have on the traveling public. Also imagine the value that they would be to us in the neater and nicer method of serving food and looking out for the passengers' welfare.

I'm not suggesting at all the flapper type of girl. You know nurses as well as I do, and you know that they are not given to flightiness—I mean in the head. The average graduate nurse is a girl with some horse sense and is very practical.

The young women that we would select would naturally be intelligent and could handle what traffic work aboard was necessary, such as keeping of records, filling out reports, issuing tickets, etc., etc. They would probably do this as well or better than the average young fellow. Further, we admit to ourselves that we are going to train couriers for ultimate jobs ashore in various traffic capacities, and we know between ourselves that there is anything but a dearth of opportunities in sight.

As to the qualifications of the proposed young women couriers, their first paramount qualification would be that of a graduate nurse and secondly, young women who have been around and are familiar with general travel— rail, steamer or air. Such young women are available here.

This is just a passing thought and I want to pass it on to you.

On May 15, 1930, members of the flight crew, by long tradition an all-male team, flying on United Air Lines California– Chicago route found an intruder in their midst: a woman.

The interloper was Ellen Church, a young nurse who had been selected by the airline to try out a brand-new job. She was to be a stewardess. She turned out to be such a useful and pleasant addition to the rough world of flight that she was asked to select and train seven other women to serve with her.

At first, some of the pilots took the whole idea of stewardesses as kind of a joke [as Ellen Church later recalled]. Then they realized that they didn't have to hand out box lunches and take care of sick passengers any more.

I don't think I made many flights when we went straight through. I had an awfully hard time getting across Wendover on the Nevada line. We always seemed to run into bad weather there, and we had to sit it out on the ground with no accommodations. There was a little weather station there, where the passengers got some space, but they usually slept on horse blankets on the floor.

The nearest place to the landing strip where the passengers could eat was the railroad station but you had to walk through the mud to get there and then walk back again. Sometimes it was necessary to make arrangements for passengers to go by train to their destination. That was our responsibility too.

We spent more time with the passengers than the girls do today. It was not like today, for we went slower and had more time to show people points of interest.

There was much more air sickness than there is now, and ventilation was pretty crude. There were heated planes, but there must have been leakage in some of them, leakage of the exhaust fumes. In order to rid ourselves of the fumes, we turned the heat off and then wrapped up in blankets. In looking back, I wonder why anybody rode at all.

Refueling was sometimes interesting. Sometimes we had to land in an emergency field and then we had the gas in two-and-a-half or five-gallon cans. They would form a sort of fire brigade, handing the cans from one to the other, including the stewardess and some of the passengers. Then, if we were some place where there was no crew on the field, somebody had to go out on the left wing to the engine to do something. Pilot and co-pilot were busy inside, so the third member of the crew had to go out on the wing—and that was the stewardess. We did it without murmur because of the argument that when a third person was needed for something like that the third person should be a man.

The first stewardesses hired by American Airlines got a grand total of three days of instruction. The "course" consisted of a trip to the Curtiss Condor factory for a quick look at the airplane, some very sketchy lessons on how to treat passengers aboard and finally one familiarization ride aboard an American Airlines ship. The women were to be no taller than five feet four inches for a very good reason: that was the height of the aircraft passenger cabin in which they were to work. In the early days, the job of airline stewardess became one of the most glamorous in the world as these trim young ladies in their snappy military uniforms began landing and taking off at airports around the country. Soon every little girl wanted to grow up to be an airline stewardess.

Having conquered the United States, American aviation experts began thinking of conquering the oceans as well. In the mid-twenties a young Yale graduate named Juan Trippe, with grand ideas in the world of aviation, founded an airline that opened operations between Key West and Cuba. As Pan American World Airways, it pushed through the islands into Central and South America. By the early 1930s, however, Pan American was looking at the Pacific and Atlantic, contemplating regularly scheduled flights across these vast ocean wastes.

Trippe invested millions in the development of several long-range seaplanes designed exclusively for this cross-ocean travel. "A seaplane carries its own airport on its bottom," he said.

His first goal was to leap the vast Pacific. He proposed a regularly scheduled airline operating between San Francisco, Honolulu and Manila—a route of 8200 miles, all told, with one tremendous nonstop ocean crossing of some 2400 miles, from California to Hawaii, a distance longer than the hop from Newfoundland to Ireland. Pilots who dared to fly this distance were still given banquets and tickertape parades up Broadway. Yet here was an American businessman suggesting that the same distance be flown by commercial transport on a regularly scheduled basis.

Modern aviation technology worked up to achieve Pan Am's unbelievable goals. The science of aerial navigation was being perfected, a combination of instrument-flying and celestial navigation. Pan American assigned crews of five men—two pilots, a radio officer, a navigator, and a mechanic—training them to negotiate ocean flights together. Meteorology, the science of forecasting weather, was also to play an important part in these ocean crossings. Radio relays made it possible for a pilot thousands of miles from land to get in touch with ground stations, checking current position and the weather ahead comfortably and swiftly.

However, the Pacific was not to be crossed merely on radio beams, teamwork, weather reports and a few star sights. Pan

American construction crews had to build relay stations that would guide and service the planes as they leaped in stages across each lap. In March of 1935, the 8000-ton chartered steamer *North Haven* was loaded in Seattle, Washington, then moved to San Francisco, California, to take on more materials and machinery designed to lay Pan American's steppingstones to the Orient. On board were 109,000 items of equipment: enough lumber to construct several villages complete with hotels, generating and refrigerating machinery, tractors, food, clothing, not to mention 120 construction laborers, engineers and flyers.

Four days out of Hawaii, the ship reached the lonely island of Midway. Here the crews found a protected lagoon, perfect for a seaplane landing. Sixty men debarked at Midway to begin work while the rest of the expedition sailed on to Wake to the west. Wake was reached by April's end and supplies were unloaded through the surf. A dock, a road, a 200-yard railway and a modern frontier village was soon erected.

On sailed the *North Haven* to Guam, a relatively civilized island already inhabited by 18,000 people. Pan American crews worked with the local Marine Corps garrison to establish the third flying base.

When the big day came the China Clipper, about to make its first ocean flight, was surely the most famous plane of its time. As it stood poised for its 8210-mile leap, President Roosevelt said, "I thrill to the wonder of it all." By two in the afternoon of November 22, 1935, close to 200,000 people had assembled along the banks of San Francisco Bay to watch the majestic takeoff of the giant four-engine Martin flying boat, a ceremony recorded over an international radio hookup.

Mr. Trippe stepped to a microphone and asked over the radio, "China Clipper, are you ready?" The dramatic reply crackled back, "China Clipper, Captain Musick, standing by for orders, sir." President Trippe spoke again. Turning toward the plane, he commanded, "Captain Musick, you have your sailing orders. Cast off and depart for Manila in accordance therewith."

The China Clipper lumbered out into the great harbor, foam splashing from its sides. It roared past hundreds of thousands of cheering people, skimmed over whistling boats, its giant engines thundering, climbed past the unfinished Golden Gate Bridge, circled in a last farewell to the crowd, and then soared westward out of sight.

The China Clipper, a Martin M-130, had a wing span only 15 feet less than today's Boeing 707s, but it weighed a sixth as much, flew less than a fourth as fast, could carry a hundred fewer passengers, cost $417,201 (compared to $6,757,000, the price of a new 707 in 1958). But the China Clipper was perhaps the roomiest airplane ever made. It had upper berths and lower berths, tables for six and tables for two, big lounges with long settees. It was indeed part ship and part plane. When it let down in the waters of Pearl Harbor less than a day out of California, it captured the principal headline from a volcanic eruption of Mauna Loa, news in Hawaii that could push the outcome of a presidential election to the bottom of the front page. Honolulu's *Star Bulletin's* streamer head shouted:

PACIFIC AIRMAIL SERVICE OPENS!
SILVER CHINA CLIPPER
ARRIVES FROM ALAMEDA IN 21½ HOURS!

On Sunday the twenty-fourth, the day after it landed, there was a rush to load the Clipper for its run onward to Wake, Guam and the Philippines. Twelve crates of turkeys came aboard, followed by cranberries, sweet potatoes and mincemeat. The colonies on Wake and Guam were about to celebrate the first Thanksgiving on those far islands. China Clipper got to Manila sixty hours after it had left San Francisco. The Pacific had been spanned, and the next year regular service would begin. Pan Am turned its attention to the Atlantic. Even with the improvements made up to 1939 there was no way to get across the Atlantic without refueling. For five years Pan Am's meteorologists had made special studies of the Atlantic weather and for two years they had prepared daily transatlantic weather maps. The leap from Bermuda to the Azores is 2020 miles, from New York to the Azores 2397 miles, both of them comparable with the longest of the Pacific leaps, San

Francisco–Honolulu. But there were other problems: seasonal sea swells in the Azores, possible harbor ice at every eastern seaboard port as far south as South Carolina.

Pan American opened its service in late May of 1939 and by fall of 1941 it had completed 433 scheduled flights flying varied routes: New York–Horta–Lisbon–Marseilles, New York –Newfoundland–Ireland–Southampton–New York–Bermuda– Lisbon. And in reverse, Lisbon–Bolama–Belém–Trinidad– Puerto Rico–New York. Sometimes it had to land as far south as Miami to escape the ice of the eastern U.S. seaboard.

The service was an immediate popular success. Air mail was ten times what the government and the airline had estimated and expected. Passenger bookings were so heavy that final lists had to be evaluated on the basis of who was more important to the national interest.

What the heavy airplanes were struggling to do—leap oceans and span continents—was being effected almost at the same time by the big airships. Germany was the most successful and standardizing the rigid airship designed by Count Zeppelin, it had begun regular commercial service as early as 1910. A company called Delag, a contraction of a typically long German company name, operated five small Zeppelins for five years before World War I, amassing a creditable record of 1588 flights carrying 34,228 passengers without injury. When the Kaiser sought to bring World War I onto British soil, the Zeppelin factory was urged to make airships with longer range, capable of carrying more armaments. Under the pressure of the conflict, the Zeppelin works made considerable headway, and immediately after the Armistice two small Zeppelins began commercial service again, carrying 2380 passengers on 103 flights without a fatality. The airships were later given to Italy and France under the reparations treaty.

Commercially the British had little luck between the wars. Its R-100, which could carry 100 passengers, did make one successful flight to Montreal in 78 hours in the summer of 1930, returning to England in 58 hours. The R-101, however, crashed over France the same year while en route to India. Civil airship operation was virtually nonexistent in the United States and France fared hardly better. The Russians did maintain regular service between Moscow and Sverdlovsk, but it was the Germans who ruled the airlanes along which airships flew. The *Graf Zeppelin*, which was finished in the fall of 1928, was the most successful of all. A year after it appeared, she made an historic flight from Friedrichshafen to Tokyo, covering 6980 miles. Her commander, Dr. Hugo Eckener, successor to Count Zeppelin, took her clear around the world, an odyssey that required twenty days and four hours to cover 21,255 miles. The *Graf Zeppelin* later maintained regular Atlantic service and she became a familiar if awesome sight to New York's children as she soared over Central Park. When she was finally sent out to pasture in 1937, *Graf Zeppelin* had completed 144 ocean crossings, 508 flights in all, had traveled over a million miles with 13,110 passengers. Even before she was retired the Germans had placed the mammoth, 804-foot long *Hindenburg* on the Atlantic run. Some ten times during 1936, the huge dirigible cruised the round trip between the United States and Germany, wafting though the air quietly at 78 miles per hour and offering her 50 passengers an unbelievably smooth and luxurious ride for a round trip fare of $720. This mode of transportation was so popular that close to 20 flights were scheduled for 1937 and extra cabins had been built into the sky ship for the additional passengers clamoring for berths.

Recalls her skipper:

> The *Hindenburg* was the best and finest ship for passengers. It is very regrettable that we have no airships now. On an airship you have a wonderful trip, not like an airplane where you can't see anything. It was very, very comfortable, and there was never any seasickness.
>
> The *Hindenburg* could travel at 136 kilometers [85 miles] per hour. It held seventy-two passengers. We had not only passengers, but freight and mail and everything. We even had a car and airplanes on board.
>
> We had a cabin for every passenger, hot and cold water, heating and air-conditioning. And a dining room and promenades you could walk on, as well as a lounge. There were four cooks on board. One was a pastry chef; the passengers had cake and bread every day. All the

food was fresh. And we had a bar also; the passengers could have wine and all kinds of drinks.

We were able to make telephone calls to stations in Europe and America. And we had a grand piano on board, so we held a very good concert for radio stations in America.

The ship had four Diesel engines with 1200 horsepower each. Diesels and helium—that is the safest combination for the air because the helium cannot explode. But we had no helium. We had Hitler in Germany and the Congress of the United States refused to give us any helium.

I crossed the ocean 161 times in airships. In the *Hindenburg* I made ten round trips to the States and eight to South America.

Then came the *Hindenburg's* first voyage of the year to North America, departing on May 3, 1937. The airship was held up by headwinds during the flight and thunderstorms delayed the landing. It was just after seven in the evening of May 6 when the huge Zeppelin nosed in toward the mooring at Lakehurst, New Jersey. Captain Pruss, the pilot, recounted the oft-recorded events that unfolded then:

We had bad weather. About two o'clock we were over New York, made a few circles, and then went on to Lakehurst. Then we saw a big thunderstorm over New Jersey and knew we couldn't land and thought it better to go back to the sea. We went along the coast to Atlantic City and back, and we waited for the storm to blow over to the ocean.

Then we were under the storm going to Lakehurst, and at seven o'clock we received a telegram that we could land. We went into a landing. During the landing, as the ropes were down, the explosion occurred.

I was in the control car. The explosion was under the cells of gas bags three and four in the aft. I heard a big noise and then I saw the flames. The flames were going through the whole ship and then forward to the bow. We were about 150 feet above the ground. My first idea when the explosion came was that the ropes had broken, but then I saw the flames and saw what it was. The stern fell and the bow shot up, because the gas was burning aft. I could do nothing but turn off the engines. As the control car came down to earth—we had under the control car a buffer, an air cushion—it bounced and the ship was therefore at about a height of six-and-one-half feet. The others in the crew jumped out. When the ship came down another time, I jumped out. Then the framework and all things came down and I was under the burning cells and framework. I was in the hospital in the United States about four months.

Thirty-six people were killed—a remarkably low total considering the size of the conflagration. Never again was the possibility of a commercial airship industry seriously considered among aviation experts.

The planes, for all their speed and dash, and the Zeppelins, if you count the inviting mode of conveyance *they* offered (at least before the *Hindenburg*'s demise), were hardly any competition to ocean-spanning ships. Not yet. The Clippers could take barely two dozen passengers, give or take a few, depending upon the wind, the mail load and the direction they were going. The nation loved its ships, loved to sail down to the warm waters of the West Indies, loved the midnight sailings of the great liners, loved the international crowd that mixed in the floating salons. That whoopee age of the 1920s had died hardest along the sea lanes. As the new year of 1930 ushered in a fresh decade, the cruises sailed full despite the frightful financial setback of the autumn of 1929. In some parts of America there were bread lines and apple stands, but in a few pockets, there was still a little loose money, and besides, wasn't prosperity just around the corner? The liners steamed out of New York—French, German, British, American, Swedish, Dutch, even Canadian, and edged into quays in remote way stations around the world. But the corner of

1931 was turned and there was no prosperity in sight. In 1931 and 1932, the cruise ships, at last, like everyone's pockets, were empty.

Now as the world began to recover from that first shock of the Crash, the strings loosened. Franklin Roosevelt was in the White House and happy days were here again. Prohibition was repealed, beer and whiskey flowed. The NRA flashed the sign of the blue eagle, a harbinger of the recovery to come. The WPA offered jobs. The CCC, or Civilian Conservation Corps, took the young jobless off the streets and put them to work saving the country's forests. The country began to whistle once more.

When she sailed into New York harbor in June of 1935, the *Normandie* of the French Line was the last word in floating elegance, a repository of French art, a testimonial to French taste and style, and above all else, a symbol of the return of the good life. The *Normandie* even paled the old *Ile de France*, so long a favorite with frequent travelers among the *haut monde*. Coming up the harbor at noon with all of Lower Manhattan free for lunch to shout and to wave, *Normandie* got the most tumultuous greeting ever sounded for any ship. By 1937 the number of travelers crossing the Atlantic—a figure that had dropped to 135,000 in 1934—was up to 207,000.

Cruising was once more the joy of the rich and the retired. The Hamburg-American Line's *Resolute*, the Canadian Pacific's *Empress of Britain* and the Swedish *Kungsholm* were all going around the world. Eight or ten ships sailed through the West Indies in winter, among them Hamburg-American's *Reliance*, Cunard's *Brittanic*, and the *Statendam* and *Rotterdam* of the Holland America Line. Twelve days in the Caribbean on the *Rotterdam* could cost as little as $140. Eight-day trips to Nassau and Havana from New York started at $90. The *Columbus* of the North German Lloyd Line, on cruise charter to Raymond Whitcomb, the travel agency, found its first cruise to South America so popular it ran it every year until the war. A two-month sail around South America cost $595 and up aboard the S.S. *Rotterdam* in 1937. At the ordinary luncheon passengers would sit down to face a menu with forty different items: roast beef or veal, sirloin steak or a brochette of kidneys, a cold Havana lobster, or a Jeanette of sweetbreads, a roast duck or a Jersey chicken à la Broche. Even tourist class on the ordinary summer runs of the *Statendam* would provide a choice of two kinds of entrees, cold meats and salads, fish and egg dishes, two kinds of dessert and a selection of cheeses.

For those who had been everywhere, or almost, there was the *Stella Polaris*, which, in the 1930s, set off with 100 passengers on a 112-day odyssey that would have made Homer blanch. New York to Havana through the canal and on to islands off Ecuador, down to the Galápagos, and slowly across the South Pacific to the Marquesas, to Tahiti, Bora-Bora, Pago Pago, Apia in Western Samoa, Suva in Fiji, across to New Guinea and to the unknown land of Nias on the west end of Sumatra, a stop at Cochin to see the colony of dark-skinned Jews who inhabited that Indian city, to Lorenço Marques, and home by way of St. Helena with a chance to shoot pool on Napoleon's table. That was the connoisseur's cruise.

So good was the cruise business as the 1930s drew to a close that the *Normandie's* thirty-day tour of South America in 1938 was sold out before the ads appeared in the newspapers. But feeling was mounting against the Germans, and a similar trip scheduled by the new and speedy *Bremen* in the spring of 1939 was an indication of the public's sentiment. It sailed with its crew of 800, a cruise staff of 15, and a passenger list of fewer than 300.

The great hotels along the Riviera that had been half empty in the first years of the 1930s dropped their prices and soon the faithful were returning. Bea Lillie showed up at Eden Roc in beach pajamas. Jimmie Walker, newly resigned as mayor of New York, paced the floor of his rooms at the Hôtel du Cap and did the coast with Betty Compton. Soon there was Norma Shearer and her husband, the nervous Irving Thalberg, shouting for the operator to get him Hollywood on the phone, ordering a car and chauffeur to run an errand up to Cannes in search of bicarbonate of soda. As prince of Wales and later as king, Edward Albert Christian George Andrew Patrick David swam at Eden Roc, played golf at Biarritz and romanced Mrs. Simpson at Cannes, on the Italian lakes, along the Dalmatian coast. They went to Austria for the skiing, to Budapest for the violins, to Vienna for the waltzes. They ran with an elegant wolfpack which the Archbishop of Canterbury referred to as "the odd circle" and which an old British nobleman called "that raffish group . . . that now lords it over British society, that mongrel pack of immigrant aliens . . . the most heartless and dissolute of the pleasure-loving ultra rich. . . ." Yet this group, for all the abuse heaped upon it, held a compelling fascination for the Cole Porters, for the Astors, the Goulds, and also for Doris Duke and Barbara Hutton, Pola Negri, Constance Bennett and Gloria Swanson, most of whom chose husbands from its members, gamboled with them on the beaches of the Côte d'Azur and Lido Venice, and later freed themselves in the divorce courts.

George V died and Edward was king but the romance with Mrs. Simpson didn't ebb, nor did the new king's interest in the ski lands, in yachting, and in rich but gamy friends. Finally, after considerable torment, Edward VIII stepped down and became David, Duke of Windsor. When he and Mrs. Simpson were married on the Riviera it was the romance of an age. ("Everybody needs being excited by the story of Mrs. Simpson at least once a year," Gertrude Stein said.) Spain was in the horror of a civil war, assassinations wracked Austria, the drums were rolling in Germany, Mussolini and the Blackshirts were strutting in Italy, and in the Orient Japan sent its troops into the heart of China. None of the clouds seemed to disturb the Americans, and they were back filling the ships and the Riviera hotels. Some idealists went to Spain to fight on the Loyalist side, but Franco found support in Father Coughlin, the radio priest, in the Hearst papers, *Time*, the *American Mercury* and Al Smith. Some travelers went to Italy and came back with glowing tales of how the trains were running on time under Il Duce. Posters printed in Rome said "Italy—Land of Traditional Hospitality," and the German Tourist Office was still offering a snappy little phrase that insisted "By-ways of beauty beckon you to Germany." The trains didn't run on time for Tom Wolfe. He was appalled by the tightening curbs on individual freedom, the stifling of the press, the attacks on the Jews. "We can never learn to march like these boys," he scribbled on a postcard to Maxwell Perkins, his editor at Scribner's back in New York.

There was time for one last fling, one final party. It was given by the Americans on both coasts at the same time. San Francisco began it with the Golden Gate Exposition, which opened at Treasure Island in February of 1939. Two months later New York's World's Fair at Flushing Meadows rang up its curtain, and clocked 600,000 visitors clicking through the turnstiles the first day. Its big hits were Futurama put on by General Motors, and the Aquacade staged by Billy Rose. Henri Soulé, who ran the fabulous restaurant at the French Pavilion, laid the foundations there for the ultimate unveiling of Le Pavillon, in its finest hour considered by many to be the best restaurant in America and by some the best in the world. It was a gay fair and a bright one, full of the angular shapes of modern achitecture, splashed by fountains and waterfalls tumbling from buildings, dazzled by showers of fireworks, brightened by sideshows and pretty girls. Along the Lagoon of Nations there were France with its terraced restaurant, Russia with its giant building topped by a massive steel workman, Italy, Poland, Norway and Holland—all soon to be embroiled in the dark events ahead. There was the Czechoslovakian Building left unfinished in mute but telling warning. Sixty nations had entered, all the developed world except Germany, for Hitler in the spring of 1939 was warming up in earnest. He had already overrun Czechoslovakia and his docket was full.

In August that year Edwin K. Hastings, a famed cruise director of the 1930s, later to become manager of the Waldorf and then a vice-president of Hilton Hotels, brought his last gaggle of tourists home on the *Roma*. She was due to sail from Manhattan the next day, but the Italians were showing some doubt about leaving the harbor. Hastings had dinner aboard the *Roma* that night, then strolled a short distance along the docks to have brandy with his old friend Etienne de la Garanderie, master of the *Normandie*. At the next pier stood the *Bremen*, on which he had sailed. He went aboard to see some of the ships' officers. While he nibbled cheese and drank the last of his beer, he got a nod from an officer indicating he had better leave. That signal was the only public announcement the *Bremen* had made of its intention to flee the port without further notice. Hastings scrambled up on deck, but found no gangplank. He rushed to the rail and far below at the crew entrance he saw one boarding plank, the last remaining link with the dock. He ran for it. When New York awoke next morning the *Bremen* was gone, its captain taking her inside Nantucket Lightship, guiding her over the shoals. There he cloaked himself in a convenient fog, avoiding the British cruiser *Exeter* waiting farther out to sea. The *Bremen* went north wrapped in foggy shrouds, left Iceland on its starboard bow, and ended up in Murmansk. Not so the *Columbus*, New York-bound with cruise passengers. She stopped at Havana to let off the customers, and then ran for Vera Cruz. Trapped by a British man-of-war, she scuttled herself in the Gulf of Mexico.

By September Hitler was in Poland and the war was on. Then one by one the nations began to fall. All of the Balkans, then Holland and Belgium, Norway and Denmark, and finally even France itself. The travelers were fleeing now along with the populace. Wrote Lee and Karl Erickson, two Americans trapped by the war:

> We left Senlis in a ramshackled car, bought that morning, and joined a slowly moving mass of carts, farm wagons, camions, and cars piled with bedding and families; refugees on foot, some women with thin shoes worn through, stumbling along on swollen, bleeding feet, pushing perambulators, pulling carts, carrying babies, dragging small, exhausted children by the hand. They had come from the north, from Picardy, from Flanders; they had been walking two days, three days, a week; they had eaten little, slept in fields, lying in ditches as the airplanes machine-gunned the road. They didn't know where they were going. This sluggisher of tragedy filled the road as far as the eye could see. It was moving slowly along all the lovely roads of France in the bright sunshine, like slowly flowing blood.

Soon Adolf Hitler, having turned tourist himself, would be photographed dancing a jig on the Champs de Mars in front of the Eiffel Tower. At his stand-up easel back in America, Oscar Hammerstein leaned over and wrote "The Last Time I Saw Paris." It was a loving poem and, set to melody, it became a bittersweet ballad of sadness and yet of hope that could be sung by sad and hopeful souls throughout the war years ahead.

VII | *Is This Trip Necessary? (1940-1945)*

THE FAIR THAT HAD OPENED on Flushing Meadow on a late April day in 1939 closed in September on the drum roll of war. It opened again, anticlimactically, in 1940, but before the heat of summer had waned, many of the pavilions were, like the countries that had built them, now the property of the German government.

Before the next year was done the United States was in it, too, and the war was truly worldwide. Soon the nation's young men would be exposed to more adventure than they had perhaps ever dreamed. Some were up in Attu in the Aleutians and others were on the run to Murmansk. Some would soon be in Guadalcanal, in Tarawa, in Bora-Bora, in China, Burma and India, in the Persian Gulf. Some would see Australia and Greenland, the desert of Sahara and the casbah of Casablanca. They would be in Shanghai and Vladivostok and eventually they would be in France and then in Germany.

Few Americans, even if they survived the war, would ever recover from its ramifications. This time they hadn't only seen Paree. They were seeing corners of the world hitherto restricted to wealthy travelers and adventuring writers the likes of Robert Louis Stevenson, Kipling, Lowell Thomas and John Gunther. The songs they sang reflected the labels they were collecting. "Bluebirds over the White Cliffs of Dover," "A Tulip Garden by an Old Dutch Mill," "Wake Island Woke Up Our Land," "To Be Specific It's Our Pacific," and that jingoistic catchline, "Goodbye, Mama, I'm off to Yokohama."

At home the bright cities along the coastline were urged to dim their lights because the glow was silhouetting tankers in the periscopes of German submarines. Occasionally, firing could be heard on the American shores. In Florida bits of wreckage and oil slicks floated ashore. Spies landed on Long Island and at Ponte Vedra, Florida. A Japanese submarine shelled the Oregon coast. Gas was rationed and so was liquor, and weather reports were eliminated for fear of giving helpful information to the enemy.

One by one America's great resorts closed down, converted into detention camps, barracks, hospitals and training compounds. The Jekyll Island Club, a posh retreat off the coast of Georgia, closed in the spring of 1942 after a German submarine torpedoed a tanker in St. Simon's Sound just north of the island. A Coast Guard detachment replaced the fashionable membership.

Secretary of State Cordell Hull, who had enjoyed its hospitality many times, commandeered the Greenbrier as an internment center for enemy diplomats. Later the Army bought it outright, tore out its luxurious interiors and turned it into a general hospital. Atlantic City, in the depths of a Depression-bred poverty, was conscripted too; some hotels became hospitals, others bivouacked troops during training sessions. Miami Beach became one giant command for the U.S. Army Air Corps. Clipped and pressed cadets marched up Collins Avenue, where the winter minks had paraded in happier times, and the shout of their cadence call echoed against the cement canyons that were built to store the pleasure seekers of bygone and better times. Over three hundred hotels and apartment houses were filled with troops of one variety or another. The parks became parade grounds, but there was little joy. Every night there was a strict blackout of windows that faced that ominous sea. Said a hotel employee of the time, "It was terrible, with all the ships going down out there. You couldn't even light a cigarette without holding a newspaper over it. There wasn't even a light on the ocean side of the hotels. Then I went to Chicago and you should have seen the complacency."

In Hawaii, where Matson ships had been depositing passengers at Matson Hotels in a land that widely advertised itself as "closer than you think, lovelier than you dreamed," the ships were turned to other purposes and so for that matter, were the hotels. By January 1944 there were 2 million servicemen in the Hawaiian Islands, half a million of them stationed there. So were 40,000 civilian war workers. One might have thought they considered themselves lucky, but they sang a song that went, "You can have the beach at Waikiki, and the Pali 'neath the moon, but if I ever see this rock again it'll be too goddam soon."

The Finance Department paid off the troops with U.S. currency stamped "Hawaii," a preventative in case the Japanese were to invade the islands after the bombing at Pearl. In such a case captured currency stamped "Hawaii" would not be legal tender. Prostitution was legalized and long lines snaked around the houses from 7 A.M., when they legally opened, until 1 A.M., when they legally closed. Hotel Street became a roaring canyon of shooting galleries, bars and tattoo parlors.

The permanent cadre of soldiers, sailors and Marines became known, to those who went on to fight deeper in the Pacific, as the Canefield Commandos. But snug on the Rock as they may have been, they sang a parody to "My Indiana Home," and it went:

> I've had enough of Nanikuli
> Waianae gives me a pain
> From Lualualei and Koko Bay
> Please send me home again
> I've had enough of pilikia [trouble]
> And of Kolekole Pass.
> You can take the Territory of Hawaii
> And put it under glass.

Barbed-wire strands made a decorator fringe for the few beach chairs that still remained on Waikiki. The old *Matsonia*, when it arrived in port, was dressed for battle. And the pink Royal Hawaiian, where the tourists once came? It was now a rest center for submariners who, after weeks in the deep, regrouped their spirits at 25 cents a night in the once luxurious rooms—tanning themselves on Waikiki's famous sands by day and sitting in the wicker chairs set up on the lanai with a view of the sand and the sea.

The tourist flow to Hawaii had long since stopped. Not only were there few available hotel rooms, but there were no ships. The ugly realities of the war at sea were brought home to Americans on the very first day of the war in Europe when the *Athenia,* a passenger ship of the Cunard Line with 1480 passengers and crew, was sunk off Ireland. In the ensuing two weeks over two dozen British merchant ships went to the bottom. A convoy system was hastily organized, but the German submarines toured the seas wrecking some of the most famous passenger liners that had cruised the Caribbean and sailed the Atlantic in happier days. When the Germans invaded Norway, their control of the Atlantic was even greater. As for Allied targets, the *Bremen* was eventually caught and sunk and the Italian passenger fleet laid waste. Even the great *Normandie,* snug in New York harbor, was not safe. Caught at a wharf in Manhattan when war broke out between Germany and France, she was never put to use by the French command. A week after the United States entered the conflict, she was expropriated by the U.S. Maritime Commission and earmarked for conversion into a troopship. In the early winter of 1942, while work was going on, and exactly one day after the Maritime Commission had turned her over to the U.S. Navy, a mysterious fire broke out. All available firefighting equipment was rushed to the dock. So much water was pumped into her that she capsized at her slip. She was never to sail again.

The *Normandie* had been the first of the 1000-foot liners, and had won the blue ribbon of the Atlantic. The *Queen Mary,* which regained the blue ribbon from the *Normandie* in 1938, survived the war as a lone gray wolf, carrying troops without benefit of convoy. That same year, the British launched the *Queen Elizabeth,* largest of all. She was designed to carry a crew of 1290 and 2225 passengers. Her rudder, some publicist figured out, weighed as much as the *Mayflower,* her whistles weighed a ton each, it took 60 feet of space to spread her letters across her bow, and set on end she was the same height as the Chrysler Building. She was never to carry a paying pleasure passenger until the war was over. She, too, operated as a loner, running unscathed through the sea lanes, eventually carrying 811,324 persons across 492,635 miles.

While secrecy shrouded the famous ships, the most publicized liner of the war was *Gripsholm*, which was chartered in May of 1942 to return Axis diplomats and bring American embassy and consular people back home. Painted white, with the words "DIPLOMAT SVERIGE" lettered across her hull, she sailed with all lights burning on an otherwise blacked-out sea returning the displaced, dislocated and sick to their homes.

For many Americans from inland Iowa and Kansas corn country, a trip on a troopship was the first look at the ocean, much less a crossing on an ocean liner. It was not exactly a voyage to win future friends for the shipping companies. The men lived deep in the holds, slept in stacks of bunks built to the ceiling. Many ships had two messes: breakfast that began at seven and continued until ten, and dinner that started at five in the afternoon and finished at ten at night. The lines were interminable and the traffic jams unbearable. John Steinbeck, writing of life aboard a troopship, quoted the GI tale of the miserable soldier who wormed his way up to a military policeman and pleaded, "Please, mister. Get me out of this line. I have had three breakfasts already. I ain't hungry no more. Everytime I get out of line I get shoved into another one."

As the European war pushed deeper into France and then into the Low Countries and Germany itself, American officers and soldiers were taken back into the quieter sectors for R. and R.—rest and recreation. In the early days some officers went back to London for short stays, but when the V-2 rockets began to drop on the city the "rest" often became more hazardous than duty at the front. After Paris was liberated, a number of hotels were requisitioned for R. and R. and troops began to filter back from the front for a few days of civilized urbanity. The city was tired and worn, and very expensive. There were more pedicabs pulled by bicycles, in the Asian manner, than regular cabs. But the beds were great, the food excelled the Army messes, and somehow French chefs made

powdered eggs seem palatable. The girls clustered around
Pigalle and even along the Champs Elysées. Tickets were
available for the Folies Bergère, especially if one had an extra
package of cigarettes to shove into the hole in the box-office
window. Incredibly there were still luxury goods available at
Grand Maison de Blanc. A magnificent gold brocaded table
runner with matching napkins seemed a steal at $80 to a
vacationing junior officer who had no place else to spend his
money. But no one from the forward echelons doubted that
three days in Paris was worth the ten-hour truck ride down
from the fringes of Germany.

The late Florian Niederer, renowned as a Swiss tourist
expert before the war and after it, recognized the tremendous
potential in the army of Americans being given rest in the
rear lines. He immediately organized tours of Switzerland in
a choice of resorts at a modest rate. The word of mouth
which the young Americans would bring back to their own
country after the war would be more than worth it to Switzerland. The "Kilroy Was Here" sketch, with its long-nosed, bald-
headed chap poking his face over walls, was popular then,
and Niederer called his Swiss trips Kilroy Tours. They were
immensely popular with the GIs, who considered them the
plums of all the R. and R. trips.

Perhaps the most elaborate resort of all was the Riviera
Rest Command, which was set up along the famed French
Côte d'Azur as soon as the war had pressed onward. Enlisted
men were given all of Nice. Male officers were in Cannes and
nurses and WACs were awarded Juan-les-Pins, the once elaborate resort just below Cap d'Antibes. General officers were quar-
tered in the superelegant Hôtel du Cap. It was a promotion
for the famed resort, which during World War I had been
used as a rest billet for Army nurses.

When the Germans invaded France during the Second
World War the Sella family closed the hotel, hid all the silver
and the wine in the mountains and retreated to a small villa
in Grasse. Not very long after the Americans landed along the
Riviera beaches in the summer of 1944, Sella got word that
Eisenhower was coming to the hotel for a reunion with seventeen generals. Sella had one bicycle and no staff and there were

650 mines in the gardens. By the next day his wine and silver had been retrieved from its hiding place and a hundred Army engineers were removing the mines. Although a plan of the minefields had been found, two Americans were wounded clearing the grounds. Eisenhower spent two nights and then returned to the pursuit of the war. Before the Hôtel du Cap went back on a civilian basis, 135 generals had stayed there to get their own rest.

Along the Riviera, there was one overriding rule. No saluting. Nice was off limits to officers, but the traffic between Cannes and Juan-les-Pins was heavy. The lady officers held a buffet every night at Les Ambassadeurs in Juan-les-Pins. The Palm Beach Casino became a huge Army nightclub. If you made a reservation before eleven in the morning, you could have dinner there with your date free instead of eating at your hotel. Prices for drinks at the Army's bars were controlled at about 40 cents, but the regular commercial bars began early to establish their reputation for high living at high cost. A martini at the Carlton Hotel cost $2, but it was lush even then. As one officer wrote home at the time, "The bar is always full of French civilians so smartly dressed and smelling so good, and the dining room at night was always full of couples —officers and French girls, officers and nurses, and American service couples on their honeymoon."

American Express had already arrived on the scene and tours were organized with sight-seeing guides pointing out the villas owned by Chevalier, Maxine Elliott and Frank Gould, the American who had built Juan-les-Pins. The buses stopped for lunch at the elegant Château Madrid, a castle high in the hills, and skipped around Monte Carlo, which was off limits to American personnel. The local kids traded coins and casino chips for cigarettes, the best currency for bargaining. At the local rate of exchange, seven cigarettes were worth one Monte Carlo chip, five were equal to four coins. Even in the best perfume shops, where prices were eye-popping, a carton of cigarettes was worth $15 against the list price, and most shopkeepers were equally ready to exchange $15 worth of overpriced French merchandise for one pair of pajamas bought in the Post Exchange for $1.60. Soap was worth $1.00 a cake, chocolate bars were 50 cents, but business was also transacted in shoes, uniform trousers and bathrobes. A maid in the Provençal Hotel in Juan-les-Pins offered one Red Cross girl $200 worth of francs for a two-year-old tweed coat from Saks.

As the Allies pressed across Germany, bombing the routes ahead of them, famous hotels and resorts surrendered to the war and crumbled in flames. One that was spared was the famed retreat of Wiesbaden with its baths and gaming parlors. It was quickly requisitioned by the U.S. Army Air Corps. Members of the Allied land forces, conditioned to winning little more than piles of rubble, were astounded that Wiesbaden had remained so clear of damage. It was the scuttlebutt of the day that the Air Corps bombers had carefully preserved it as a comfortable nest that would be ready for them when they moved onward into the Reich.

In beleaguered Berlin, the old Habel Hotel disappeared into dust, then the Bristol and finally the Kaiserhof, an old-fashioned house that had been a favorite of Hitler's. The celebrated Adlon at Number 1 Unter den Linden, where so many Americans had come, still stood. Somewhere in its old registers were the names of Teddy Roosevelt, Franklin Roosevelt, Mary Pickford, Doris Duke, Herbert Hoover—so many Americans that it was known as America House Unter den Linden. Hedda and Louis Adlon, who had run it from 1922 until 1945, fled to the family house at Neufahrland when the end neared. They survived the waves of Russians as they came, but one day a servant referred to Louis as "Generaldirektor," and the Russians, thinking they had found a high Nazi officer, dragged him away. Hedda followed his trail for nearly two weeks and found him at last dead in a mound of earth. From an old woman she bought a sheet for his shroud, and from an aged man who was knocking together a hut for shelter she bought a few extra boards and that was his coffin. As the zones were later marked out, old Louis lay outside the city, in the Russian zone. Years later Hedda had him moved to Berlin and bought him a linden tree. "He loved so much the linden," she said. The Adlon was gone, too. Although it had survived the Battle of Berlin, it had gone up in flames the night the Armistice was signed. Its steel was taken to help build the Stalinallee and only 100 rooms, in what had been the old couriers' wing, were still operating. The site, which had personally been chosen by Kaiser Wilhelm II, was, worst of all, just beyond the Brandenburg Gate, inside Soviet-controlled East Germany.

The Americans returned home at last, sailing into San Francisco and into New York Harbor to happy welcomes that were repeated again and again, long after the ships with the first returning troops had arrived. To the homebound men the snags often seemed preposterous. There were ships that were left in mid-Hudson overnight because they had come up the harbor too late in the afternoon to be unloaded by the dockhands. There were the officious garrison officers who snapped out orders to the men and officers just off the ships from duty overseas. And then there were the tales of the difficulties of gas rationing, the unavailability of new cars and other hardships endured by civilians. America's fighting force had seen the awfulness of war, but it had also seen the world as no invaders had viewed it before. Paris was familiar, but so was Brisbane. Tokyo was an experience not to be forgotten, Switzerland was a place to which all who had seen it vowed to return. And what of India and Morocco, England with its warm memories of early war? It was these experiences, told and retold a hundred times to those who had stayed at home, that built up such an interest and such a thirst to see the world. To these origins could be traced the beginnings of an era of tourist migrations such as the world had never known.

In 1930, the year Boeing Air Transport first suggested putting stewardesses on airplanes, Western Airlines unveiled the world's largest land plane, a four-engine Fokker-32 which advertised its size with a chorus line dancing on a wing. Only five years later, Pan Am sent its China Clipper from San Francisco clear to Manila, a flight that took 60 hours. Said President Roosevelt, "I thrill to the wonder of it all."

Resorts vied to invite celebrities. Albert Einstein examined the wonders of Palm Springs, California, while Sun Valley, in Idaho, attracted Gary Cooper and Hemingway.

Germany's dirigibles had all the comforts of ocean liners. The *Graf Zeppelin*, which appeared in 1928, flew from Germany to Tokyo, flying nearly 7,000 miles. Then her commander took her clear around the world. She made regular Atlantic crossings after that, traveling over a million miles without mishap. The *Hindenburg*, her successor, flew quietly at 78 miles an hour, carrying 50 passengers who paid $720 round trip.

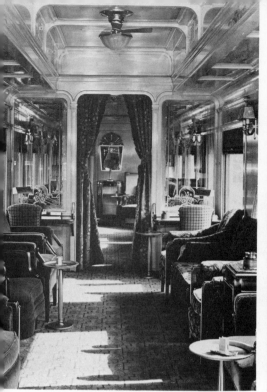

UNION PACIFIC RAILROAD   AMERICAN AIRLINES

Trains of the Thirties were comfortable and inviting. The Colorado Car of the Union Pacific's *Columbine*, above, had writing desks, overstuffed chairs and a parlor look. Early stewardesses on American Airlines were all registered nurses. Here they are lined up before a "Silent Condor." Some stewardesses even dressed as nurses, below.

BOEING

The era after World War II still belonged to the great liners. This former German ship was given to France as reparations and became the *Liberté*, redolent of midnight sailings and transatlantic chic. The French Line had lost most of its fleet.

The first glamour airplane of the postwar period was Lockheed's sleek Constellation, which could fly at 300 miles an hour and had a pressurized cabin. TWA sent the first one to Europe in 1946, but by 1947 it had extended its routes to Egypt and planes were flying over the ancient pyramids. It took another three decades before Americans could make the ultimate trip to China to see the Yangtze Bridge in Nanking, China.

This Sitmar liner, at left, like other ships, became not a carrier but a floating resort. New ships will have standard-size rooms, spend less time at sea burning fuel. Rakish train, below, developed in Germany, operates by magnetic levitation. It speeds at 250 mph, city-center to city-center.

Tomorrow's traveler abroad in Los Angeles will be able to roll from the glass towers of Western's Bonaventure Hotel to the theater aboard an elevated People Mover, which the artist has superimposed on the photo above. Hotels of the future, such as the rendering, below, of the Sharjah Inter-Continental in the United Arab Emirates, will have a new look, new computerized business techniques and a new role as an oasis.

In the new world all things change. Who could have imagined that New York's venerable Commodore Hotel would emerge as a glass box called the Grand Hyatt with a glass cage hanging over Forty-second Street? Recalling the bouncing stagecoaches, who could have dreamed of a space shuttle that may well open the way to interplanetary travel?

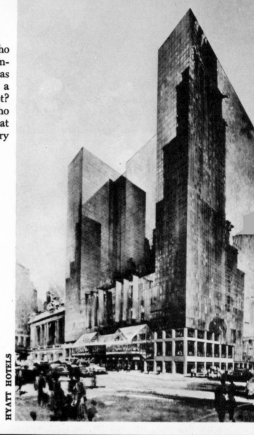

HYATT HOTELS

NASA

VIII | *Peace and Peregrination (1946-1957)*

THE FIRST SPRING WITHOUT WAR blossomed in a burst of dogwood, azalea and pink apple blossoms. It was 1946.

The travel agents and innkeepers had hardly wasted a minute. This was "Victory Vacation Year," and the posters urged everybody to go somewhere and take a rest. "You've earned it—Now enjoy it," said the slogan. The enjoyment was scheduled to start officially with the Memorial Day weekend, a somber one for many an American family. Still, all the leading hotels in Pennsylvania's Poconos were reporting sellout business. In Chicago, the Georgian Bay Line was sending its *North American* to Mackinac Island; Goderich, Ontario; Escanaba and Sturgeon Bay. In Detroit, the *South American* would sail on its Memorial Day cruise bound for the Thousand Islands with stops at Cleveland, Ogdensburg and Toronto. The old Hudson Day Line was back on its river and would embark daily beginning May 25 for Newburgh, Poughkeepsie, Kingston Point, Catskill and Hudson. They might not have seemed like exotic ports to those who had been at Bora Bora or Juan-les-Pins, but to the folks who hadn't been anywhere for four years it all seemed entrancing. And besides, all was right with the world, for moonlight sailing up the Hudson was to begin June 1.

A glimmer of the gay cruise days before the war appeared. The old *Stella Polaris*, famous in the 1930s for her North Cape cruises from England to Scandinavia and her super luxury round-the-world excursions, was back, captured intact from the

213

Germans. She had spent the war first as a floating clubhouse for Nazi submarine crews, and later as a dockside hotel for German occupation troops in Norway. Now flying the Norwegian flag, the ship was returning to the Caribbean, where American Express had scheduled tours at half a dozen ports and a whole weekend in Havana. The same week, the Grace Line brought the brand new Santa Barbara into New York, ready to sail for South America, as the first postwar passenger ship flying the American flag. The first dream of the new era, she had beds that would unfold at the press of a button, partitions that would disappear to make suites. There were windows fitted with Venetian blinds instead of the old unworkable portholes. Each room had individually controlled air conditioning. Her 52 passengers could swim in an open-air, tiled swimming pool, loll in the sun on a cork beach, watch movies at night from a covered veranda. Compactness was the byword. The passengers on the maiden voyage would sail in company with newsprint and noodles for Panama, textiles and raisins for Bonaventura and refrigerators and flashlight batteries for Peru.

Eight years and three weeks from the day she was launched, the giant *Queen Elizabeth*, the largest ship afloat, left Britain on her maiden voyage as a luxury liner. She had been launched at the height of the Munich crisis, and then in 1940, with some of her launching gear still attached, she moved down the Clyde River seemingly bound for Southampton. With a skeleton crew aboard and no passengers, the ship that had never been to sea slipped by the German scout planes and showed up in New York Harbor five days later. From there she sailed clear to Singapore to be outfitted as a troopship. She could take a complete division and all its gear at one swallow. Among the 811,000 troops she had carried during the war were parts of the American 8th, 44th, 70th and 104th Divisions, and the ragged remnants of Hitler's vaunted Afrika Corps brought to prison camps in America.

Now in her dress suit at last, she had been given new turbines, outfitted in all cabins with special colorless glass made in Czechoslovakia. Her galleys, which had prepared 2 million meals in wartime, were revamped for high cuisine. Her original wine cellar, which had lived through the war safe in an English vault, was returned to her complete with vintages that were now unobtainable. Now, as fireboats sprayed plumes of water around her, the big QE sailed into New York ready to take 2314 paying passengers that hopefully would earn badly needed dollar credits for Britain.

The decade was perhaps still to belong, as had those before the war, to the great passenger liners. But the conflict had fostered monumental developments and they were to unfold in a dramatic rush in the ten years ahead.

Four months before the war broke out in September of 1939, Pan American had opened a regular flying schedule between New York and Europe. To maintain the service its lumbering Clippers had to chart short leapfrog routes, stopping in the Azores and Lisbon on the way to Marseilles or putting in at Newfoundland before skimming into the harbor at Southampton. The western route back to the New World was even more cumbersome. Headwinds forced the navigators to take puddle-jumping crossings, swinging as far south as Puerto Rico. Ice in the winter months often required the flying boats bound for New York to land as far away as Miami.

The war did away with all that. The most dramatic outgrowth of the six-year fight was the emergence of the land plane as a quick, dependable means of long-distance transport, both cross-country and transatlantic. The domestic air system had been forced to grow up in a hurry. The few four-engine planes that had existed at war's outbreak had proliferated, at war's end, into a vast worldwide network. The Air Transport Command had forced a transatlantic ferry route by way of Natal in Brazil to Dakar in Africa. It carried high-priority war personnel and war correspondents, in the closing days, from New York clear to London and Paris. Over 800,000 wounded and sick had been flown home from all the world's theaters. By 1945 more scheduled flights were churning over the world's oceans than had existed between Chicago and New York when the war began.

Now with the peace, the huge bombers and mammoth transport planes were quickly redesigned for commercial use. The revolutionary Constellation, thin and sleek, with a slight curve in its long fuselage that gave it the appearance of a diet-slim porpoise, made its first appearance in the early days of 1943. Now it would be the fast glamour plane of the first postwar period. Its cabin was pressurized and it could cruise at 300 miles an hour. The old DC-3, which had worked its way through the war with such dependable success, would still be in the skies for a long time, but a new and larger version was out, the DC-4. It wasn't pressurized, but it was bigger and faster and, what's more, it was on hand. Pan American alone had 71 of them in its fleet. Racing to reestablish its transatlantic routes, Pan American was first off the blocks, sending a DC-4 to London in January of 1945. It got its passengers, most of them businessmen, to England in 15 hours and 50 minutes. All the world gasped. American Airlines, a domestic carrier, took a look at the transatlantic market, found the northern route open and, establishing a subsidiary company called American Overseas Airlines, sent its first group of passengers across the ocean to Scandinavia in October of 1945. TWA, with its distinctive triple-tailed Constellations, finally got its first flight off to Europe in early February of 1946. It was late, but it had a lot to offer: the fast Connie had a pressurized cabin that would allow the pilot to fly above the weather without popping an eardrum or causing a loss of oxygen in the cabin. TWA's first transatlantic Lockheed Constellation 049 took off on a cold, windy day from La Guardia field. Aboard were American and French businessmen, two fashion designers, government observers and an ex-GI off to bring home his Belgian war bride. They each paid $375 for a one-way passage. The first flight was delayed for eleven minutes because a passenger had misplaced her handbag, but at 2:11 Captain Harold Blackburn, a famous pilot, later to be the subject of a *New Yorker* profile, took her into the air and headed out along the Lindbergh route by way of Gander, Newfoundland and Shannon. He brought her down at Orly Airport the next afternoon, having covered the 3700-mile run in 19 hours and 46 minutes, including time for refueling. TWA now began to trumpet in its ads that it had "put the U.S. within swift, easy reach of Paris." It guaranteed to make the crossing in 22 hours using its Connies and Sky Masters. Service was scheduled for once a week.

The United States was not long alone on the airways. The foreign carriers, which had disappeared along with their countries, suddenly sprang into being, among them Air France; KLM; Sabena, the Belgian line; and SAS. Swissair soon flew across the Atlantic bringing an unusual comfort to its unpressurized DC-4s. Eventually there would be the Italians, the Spaniards and ultimately even Lufthansa, the German line, would be resuscitated. The European lines strove for luxury. By 1947 Air France, seeking the same aura of luxury in the air for which it had long been celebrated on the sea, had begun a twice-a-week nonstop service between New York and Paris on a new Constellation. Aboard this Golden Comet service, as it was called, the cabin load was limited to 18 passengers, who paid $370 for passage plus $125 for a berth. The select group was served a magnificent 440-mile meal so called because the plane had traveled 440 miles in the time it took to serve it. Pâté de foie gras, lobster salad, breast of chicken with truffles and rice, a split of champagne, a separate serving of asparagus, an assortment of cheeses and fruit, ice cream and chocolate sauce, brandy and coffee all followed, setting the style for elaborate meals that all airlines were soon to adopt on the competitive North Atlantic run. The trip took an incredible 12 hours. BOAC, Sabena, Pan Am, AOA and TWA were all scheduling 16 to 20 hours including stops at Gander, Shannon or Prestwick. Pan Am, in the first days, had to transfer its Paris-bound passengers to Panair do Brasil in London, and it therefore wasn't even in the Paris running.

The Air France flights to Paris marked the beginning of a new era in speed and luxurious travel, which had never been seen before and would soon disappear, perhaps never to be seen again. Pan American and other airlines later introduced the lumbering old Boeing Stratocruisers, perhaps the most comfortable land plane yet designed. The huge Boeings, really converted bombers, were fitted with big stretch-out seats and berths on the main deck, and a cocktail lounge built into the deck below. On the 2600-mile flight from Hawaii to Los Angeles, passengers could relax in their seats with their legs stretched out, go below for a drink, stay for dinner in the snug lounge and then later climb upstairs to their berths. The lights of Los Angeles would twinkle in the first light of dawn. There would just be time for coffee before the landing.

Soon came the DC-6, a larger, improved and pressurized version of the DC-4. On SAS, a passenger could board a plane in Stockholm and for $50 extra sleep in a berth all the way to New York. TWA fitted berths into Constellations, even used them on the midnight run to California, a favorite flight of the stars, who nibbled at the flying buffet table, then went to bed until home sweet Hollywood came up under the wings.

Although the feats of the transatlantic planes were the most dramatic achievement of the first postwar years, the ships, as they had before the war, carried most of the passengers. The passenger fleets of most countries had been heavily hit, but Britain had been lucky. Besides the *Queen Elizabeth*, the Cunard Line also had the *Queen Mary*, and the *Mauretania* which had made its maiden voyage in June 1939, just a few months before war broke out. It also pressed into passenger service the elderly *Aquitania*, which had been a debutante in World War I. It had made its maiden voyage on May 30, 1914. The *Queen Mary*, first ship to exceed 1000 feet, had been launched in 1934, and in 1938 had recaptured the blue ribbon of the Atlantic from the *Normandie*. Soon Cunard would produce the *Caronia*, a luxury ship built expressly for cruises. Painted a pastel green, she became known as the Green Goddess, a sobriquet (and a curious color) that did not prevent her from being a popular vehicle for America's widowed blue-haired ladies on their worldwide excursions.

The French Line, which had lost two thirds of its fleet along with its pride, the beautiful *Normandie*, was in sad condition. For the first years it limped along with the aging *de Grasse*, which carried 711 passengers in two classes, and the tiny *Wisconsin* and *Oregon*, neither of which had room for a hundred trippers.

Eventually the *Ile de France*, so redolent of midnight sailings and other warm memories of the years before the war, came back. From the Germans the French got the old *Europa*, rechristened her the *Liberté*. But before she could be put in service, she broke from her moorings at Le Havre, stumbled into the old *Paris*, which was lying on the bottom, and sank, too. Finally refloated, she became what was probably at the time the best French restaurant in the world. Her dining salon sparkled with the old élan. Spun sugar baskets full of rich petits fours paraded from its bakeshop. Crêpes suzette broke out like brush fires in the first-class dining salon, and the caviar thumped onto chilled plates in huge pearl-gray globs.

The Italians, whose best ships were all resting on the bottom, would have to make do with the old *Saturnia* and *Vulcania* for years to come. The Dutch, their fleet torn to shreds, rushed to recover the Atlantic trade. Their first efforts sought not the luxury market, but the tourist-class passengers. Lifting the poor misbegotten soul out of steerage, the Holland-America Line produced the *Ryndam* and the *Maasdam*, a pair of ships almost totally devoted to tourist-class passengers. Limited accommodations were made available for a few passengers in first class, but that was merely to qualify the vessels as two-class ships. The *Nieuw Amsterdam*, which was on a South American cruise when the Germans invaded Holland, returned to New York, debarked her 700 tourists and made ready to join the fight. She had carried 8000 troops at a time on numerous voyages from the Pacific Coast to Australia, finally returning to Rotterdam a year after the Germans surrendered. Now she was back in gay colors as a luxury liner.

Even as the first travelers edged gingerly toward the docks and the airports, the first guidebooks began to appear. In the eight years or so that had elapsed since bona fide tourists were tramping the drafty monuments of Europe with red Baedekers under their arm, travel writing had taken a sharp new turn. The first books were hardly a collection of dusty facts and figures. They were frequently gay, sometimes funny, often opinionated and uncommonly irreverent.

The venturesome travelers who went abroad in the early days often carried canned meats, chocolates, even jars of vegetables and boxes of biscuits. These provisions were to supplement the lean fare they expected to find; and of course, they would make good gifts to give the starving Europeans. The Europeans, as it turned out, were not exactly starving, and the hotels were ready with meat, fish, vegetables, eggs and even pastry. It was not to say that Europe was without problems in

the late 1940s. There was still rationing and the first travelers
had to carry ration coupons in France. "Café national," as it was
so quaintly called, was not exactly made from coffee grounds,
but it was hot. Often, alas, the hotel rooms were not. Coal was
still in short supply. Sugar was not available on trains, in small
cafés and off-the-line cities. The Communists were a problem,
particularly in France and Italy, where they had won several
elected offices. Strikes were frequent. First the subways. Then
the cabdrivers. Then the electrical workers. Travelers arriving
at a depot in Paris and finding no transportation were often
obliged to hire a baggageman to push his wagon through the
streets so they could get their luggage to their hotel. Often
they had to climb the stairs to their rooms because the power
had been shut off. Still the travelers came, paying the ex-
orbitant prices with a currency that was badly in need of
devaluation.

The famed old hotels of Paris were in remarkably good
condition. The George V, built in the 1930s, was still elegant.
The French had kept it up for the high-ranking Germans who
had occupied it. The old Ritz was getting its clientele back,
the Crillon was no less creaky than it had been before the war
—after all, it had known Woodrow Wilson. The Meurice, where
a battle had raged and a machine gunner had held out on the
entresol, had been spruced up. In Deauville the du Golf and
the Normandy were back in business and the wheels were
spinning again in the casino. The Carlton in Cannes and the
Negresco in Nice had not only survived the war, but they had
lived through the years when all the Riviera was a rest center
for American troops. The Côte d'Azur had long since passed
into civilian hands now, and the flossy set was back in their
sleek cars and expensive clothes. Merle Oberon, Greta Garbo,
Sonja Henie, Doris Duke, Jack Warner and Orson Welles were
all on view. The rates were astronomical.

For Americans France had always been the first country
for a European excursion, often the only one. Italy, which had
chosen the wrong side in the war, was off to a slower start. But
now there was new word from an unexpected corner. The
British Isles, which for a century had sent travelers to the Continent in droves, were now making a bid to have other travelers

visit them. There was a worldwide dollar shortage and the British aimed to get a part of it. They had little to offer in the line of tourist amenities. Food was short and not very good. The Blitz had demolished huge chunks of London and what waited in the British hinterlands for the explorer, heaven and American Express only knew. Undaunted, the English mounted a brilliant, energetic promotional campaign. They sent London buses to America, toured them all the way to the Pacific Coast. They followed with uniformed bands, real live pearlies, pub keepers, Welsh lasses in costume—anything to make the papers and advertise Britain. Much to everyone's surprise, the drive caught on and an unmistakable trickle began. The American visitors didn't always come home ecstatic about what they had found, but Britain was earning dollars from a source it had never tapped before, and it needed them desperately. Travel abroad for its own citizens was strictly controlled.

In the best shape of all were the Swiss. The country was untouched and therefore quite as beautiful as ever. No bombs had fallen, no soldiers had gone to war. What's more, the Swiss hoteliers had gotten in a little practice with the big program of rest and recreation which they had mapped out for the Allied troops. The techniques of the Kilroy trips of wartime were applied to American civilians. All the Swiss needed was an airline. Soon they sent Swissair zooming across the Atlantic. All they could muster, even by the early 1950s, was unpressurized DC-4s, but they flew, the Swiss said, with Swiss precision. A foray into Geneva surprised the early travelers, who were busy putting up with the shortages and the distemper in France. Swiss shops were stocked with oranges, grapefruit and bananas, all premium items in France just across the border. The bread was white, the beer was rich. Incredibly, a Swiss waiter who had been tipped by a foreigner would likely stop and tell the visitor that the tip was already added into the bill. Wonder of wonders, there was soap in the bathrooms. Not the best of France's hotels could offer that.

The thin ranks of ordinary, civilian pleasure travelers were swelled considerably in the early days by the occupation forces who were sent on leave to Baden Baden, Wiesbaden, Garmisch-Partenkirchen and Berchtesgaden. The best resorts of Germany and Italy that were still in one piece were requisitioned by the occupation forces and officers, soldiers and civilian employees of the military spent luxurious leaves at prices that would never again be duplicated. Former servicemen took advantage of the GI Bill of Rights and returned to Europe, ostensibly to study. Like the expatriates who had flocked to Paris after the First World War, they hid themselves away in the garrets of the Left Bank affecting the beards and berets of French bohemia. Often they slipped so deeply into the local idiom they forgot all about their homes and families back in America. The American embassy was plagued with inquiries about boys who had disappeared into the maze of Montparnasse. Emissaries were continually exploring the back alleys of Paris in search of the missing.

The story is told of a mother in Lubbock, Texas, whose son had gone to France on the GI Bill. Letters stopped coming and she began to worry. Distraught, she finally had a chat with her mayor, who, to be helpful, passed the word on to the governor, who notified a senator in Washington. The senator, mindful of his constituency, flashed the State Department, who cabled over to the embassy in Paris. Pretty soon a vice-consul in charge of missing Americans tracked down the boy's address, took the subway to the Left Bank and after sorting out the winding streets finally came upon the house. He ascended the old groaning staircase, carefully avoiding piles of garbage and passing cats and knocked on the door of an attic room. Past the piles of dirty laundry and dirty dishes he could see three bearded bodies in striped Basque shirts and berets. He asked if any of the quaint souls happened to be Americans. One who had been cooking lunch on a gas burner admitted he was American. "Are you from Texas by any chance?" the embassy man asked. The American affirmed that he was indeed a Texan, and the man from the State Department said, "Does your name happen to be John Jones?" "Yes, I'm Jones," said the surprised beard. "Well, dammit, Jones," said the government's emissary, "why don't you write your mother?"

Foreign governments quickly opened offices in New York to promote travel abroad. Soon the bureaus spread to Chicago, and ultimately to the West Coast. The general movement was encouraged by Washington, which was eager to get dollars into the hands of dollar-short countries. The best way of implementing the Marshall Plan, many a writer wrote, was for Americans to bring over the dollars in their own hands. Colorful advertising campaigns broke in American newspapers and magazines. Editors searched to find editorial material to pour around the ads. Newspapers and magazines began to hire travel editors. In the old days the travel editor had worn three hats, often being charged with pasting together the radio listings and writing the real-estate puffs on the side. Many newspapers and magazines recognized a new breed, a writer totally dedicated to reporting travel as a full-time beat. Some travel writers became syndicated columnists, publications issued whole sections, sometimes whole issues, devoted to travel. While the pap and the friendly story that would please the advertiser continued to pour from many a Smith-Corona, some travel writers became reflective critics of the scene, evaluating places, practices, planes, inns and ships in the manner of the arbiters of the movies, theater, art and the emerging world of television.

While Americans were enjoined to journey off to Europe even in the fledgling days of its convalescence, enticed into cruising aboard the *Caronia*, Cunard's 1949 entry in the world cruise-ship sweepstakes, and even invited to fly off to the West Indies on island-hopping trips, America's homegrown tourist industry was barely stirring. The railroads seemed exhausted from their wartime efforts. The Chesapeake and Ohio had a brief fling introducing movies aboard its trains, but beyond that the eastern railroads scarcely bothered to clean their sooty windows. In the West, the trainmen were conscious of the scenery and several made bids for the domestic traveler by equipping their trains with double-decker cars surmounted with Vista Domes.

Automobiles were at a premium in the first days of peace, but Detroit ultimately retooled itself and the pleasure wagons once more rolled down the assembly lines, this time at an ever accelerating speed. Soon long strings of cars were crowding the roads and war veterans who had marveled at the autobahns of

Germany were wondering aloud why we hadn't long since built the same thing. Finally, in the early 1950s, the New Jersey Turnpike opened, spreading more than a hundred miles of superhighway from Delaware north almost to the George Washington Bridge. The turnpike broke the jam of cars entering New York City from the south and millions of Americans were encouraged to wheel out the car for trips. The New York State Thruway threaded more than 400 miles from New York City to Albany, then west, following the route of the Erie Canal. Inspired, Ohio, Indiana, Kansas and states all across the nation laid down superhighways. There began a renaissance of automobile travel in the United States which spilled over into Canada and even Alaska, where the Army had in wartime rushed to complete the Alcan Highway. A midwar emergency measure, it was now a tourist trail.

The new rush to the roads marked the demise of the railroads, especially the eastern railroads, as a means of vacation travel. Although the European trains and those in Japan created new comforts and new speeds on the rails, America's train systems remained old-fashioned, unimaginative and even indifferent to the pleasure travelers. As people sped off in rakish new cars to roll down broad new highways, a new world of motel living emerged. The shabby cabins by the side of the road gave way to long low buildings with modern rooms, air conditioning, television, swimming pools and eventually such gimcracks as ice-cube makers, coffee machines in every room and a bed that would give its occupant a massage. Some motels added gift shops and restaurants and later even nightclubs. The motel forced the aging midcity hotels to modify their rates, spruce up their interiors and provide new enticements for the car traveler. Once relegated to the city limits on the edge of town, the motels began to move into the middle of cities. Ultimately they invaded the resorts.

Many of the nation's big hotels had done war duty as detention centers for foreign diplomats, hospitals and rest centers. Now, as the military returned the property, a vast program of refurbishment began. The famous old Greenbrier in West Virginia brought in Dorothy Draper and the venerable halls bloomed with cabbage roses on the walls, green carpets on the floors, and flowered prints and gold-framed pictures. The Royal Hawaiian, its submariners gone, got new furniture and a new coat of pink paint and opened its doors once more. Palm Beach glittered in a new season as early as the winter of 1945–1946. Miami Beach, which had opened forty new hotels in 1940 only to find itself a part of the war effort the following year, redecorated and girded itself for a new boom. It wasn't long in coming.

What really put over Miami Beach and made it a haven for commuters from the cold, and a year-round resort as well, was a new mode of transportation: the airplane. Rail travel took an agonizing twenty-four hours from New York, and to be comfortable en route cost a lot of money. Car travel was a tedious trip. The superhighways hadn't come to the South, which still enjoyed catching the Yankees in speed traps as they moved through the towns. Burly sheriffs and constables with blackjacks lying aside them, often with their children licking lollypops on the back seat, made it a habit, if not a despicable practice, to stop out-of-state cars for all sorts of infractions. The fine was often collected on the spot. The alternative was a day or two in the backwater town until the case came up. It all reached a fine art in the metropolis of Ludovici, Georgia, where a lookout in a second story hideaway of a midtown building pushed a button turning the traffic light to red just as motorists—out-of-state motorists—were beneath it. A constable waiting by the side of the road waved the traveler to the side and announced the charge. The case would come up in three days. The verdict was predictable. The alternative: $15 cash on the spot.

The new airplane flew over all that. They got you there in four or five hours instead of four or five days on the road or twenty-four hours on the rails. Miami Beach looked at the new Riviera summer season and thought about one of its own. National Airlines came up with Piggy Bank Vacations, offered reduced air fares packaged with reduced rates in hotels. For $19 a week, they would include a drive-yourself car. In 1946 Miami Beach gingerly added 366 new rooms in six small new hotels. Of these, the Martinique, with 137 rooms, was the largest. But the next year it started. First came the Sherry Frontenac (still small), followed by the Saxony and Sans Souci (both large). Trends were carefully watched. When the architect of the Sans Souci put in a big birdcage suspended on brass poles, the Biltmore put in eighteen. What's more, its owner didn't lose nerve like the Sans Souci and at the last minute put in plants instead of birds. It stuffed its cages with real parakeets. The new di Lido put in birdcages, too, but it added only six because it could counter with something else new: stained-glass windows in the bar.

By the season of 1954 the motels were in full bloom along northern Miami Beach. It wasn't *really* Miami Beach, the purists insisted, because it was beyond the city limits. The motel owners countered that by renting post-office boxes in Miami Beach. America was bewitched by the South Pacific and a trend to Oriental decor was sweeping the nation. The motels obliged by offering instant Orient at anywhere from $14 to $22 a day for a double room in winter and about half that in the summer. Invoking the blistering sultriness of Lamour, Michener, the Moon of Manakoora and Somerset Maugham, the motel operators festooned their inns with lanterns and gave them names like Mandalay, Tahiti and Bali. Despite the predominantly Jewish clientele in Miami, such names dear to the Arab homelands as Suez, Sahara and Tangier were also hung over doorways. The Sahara even mounted a pair of stuffed camels in the driveway. One of them had a stuffed Arab in the saddle. To provide a proper balance, one place, the Pan American, restricted itself to a gentile clientele.

The DC-7 arrived on National Airlines' Miami–New York run powered with something called a turbo-compound engine. Nobody really knew what that meant, but the bird got you to the sunlands in three and a half hours from the northern slush-lands. It beat spending six to eight days driving down and back. A lively car-hire company called Couture would, if noti-fied in advance, have one of its rentable autos waiting at the airport gate. It used mobile service stations to keep its cars oiled and gassed, offered to let its customers drop cars at other Florida points in case they wanted to do some sight-seeing in the state before they left for the North. Of the two young brothers who operated Couture, one retired in his forties and the other later became president of Hertz International.

But the motels, however informal, however attractive, however functional, however less expensive they might have been, were not to mark the end of the grand hotel. Not in Miami Beach. The next season, 1955, saw the inauguration of the supersplendid, $14 million Fontainebleau. It didn't matter that Broadway and Miami immediately called it the "Fountainbloo." It was French. Didn't it have the fleur-de-lis running up and down the jackets of the waiters in the Fleur de Lis room? Weren't its formal gardens a copy of Versailles? Didn't the rooms ($33 to $135 a day, no meals) have pictures of Paris hanging on the walls? Weren't its public washrooms labeled "messieurs" and "dames"?

For the gala opening, its diminutive New York press agent, one Leo Morrison, was as French as an inflated soufflé. He appeared in a blue beret, a white shirt and bloodshot eyes. Eh, voilà! The tricolor waved over Miami Beach. In a burst of inspiration the mayor of that other Fontainebleau, the one in France, was invited to cross the ocean. He came bearing a tree from the forest of Fontainebleau, but, alas, it failed to clear the agriculture inspectors at Idlewild Airport in New York. A quick-thinking press agent removed a Christmas tree—the season was just right—from a nursery truck, affixed it with a red, white, and blue ribbon and gave it to the mayor to plant in the Fontainebleau's front yard. The mayor planted the Christmas tree with Gallic dash. Then, laying down his spade, he looked up at the hotel, carefully surveyed its mixture of architecture and pronounced it all a giant bouillabaisse.

The Fontainebleau, with its French accents, sounded a signal all across the beach. New hotels had to have themes. Patrons at the Hotel Seville had their morning coffee in the Cafe Olé, their dinner in the Fiesta Room before moving into the Matador Room to dance. The Lucerne went Swiss. The coffee shop was named for William Tell, the dining room was called the Alpine Room, and the nightclub, the Club Chalet. The Thunderbird Motel adopted an Indian motif. Bagels and lox were served in the Bow and Arrow, drinks were passed over the bar in the Pow Wow Room and anyone en route to the rest room looked for the door marked Braves or Squaws, as the gender indicated.

When the Eden Roc Country Club and Hotel (calling it a hotel rendered it an injustice) opened the next year alongside the Fontainebleau, it plainly was unsure just where the real Eden Roc was located. Instead of summoning the place names that surround Cap d'Antibes on the French Riviera, where Eden Roc has long been a rich retreat of the smart set, it called its dining salon the Mona Lisa Room, its coffee shop became the Villa d'Este, its nightclub the Cafe Pompeii, and the saloon, Harry's American Bar. When architect Morris Lapidus, who had built both the Fontainebleau and the Eden Roc and many another Miami Beach palace, sent down a copy of the Winged Victory of Samothrace to decorate the front lawn, he immediately had a call from the Eden Roc's owner. "I spent enough money on it," the patron said with audible outrage, "so why can't I have a statue with a head?"

Each year saw the birth of at least one grand hotel known along the Beach as "this year's hotel." The loyalty of the patrons was only to newness. But south Florida weather was never sure, and now the airplanes were making it easier to explore a new frontier, the West Indies. It had started perhaps with the Caribé Hilton, a new type of hotel by a new type of hotelier. Conrad Hilton had come out of the Southwest and was expanding across the country. He had already reached into New York to buy the Waldorf-Astoria, which he had always called "the greatest of them all." Now he joined forces with the government of Puerto Rico. Using young Puerto Rican architects, the government and Hilton built a different type of hotel. All its rooms were air-conditioned, all had balconies with flaring views of the sea and the deep green of the countryside. The parched could imbibe fancy rum drinks with an unimpaired view of the ocean, even though they were seated in deep easy chairs. The bartenders had been sunk in pits. The American-style coffee shop, in an upholstered outdoor version, landed for the first time in an overseas hotel. It would service businessmen coming and going between the

States and South America, but it would offer its facilities in a resort setting.

The result was instantaneous. Puerto Rico, which had been a depressed slum pitted with shanty towns and open sewage, found new courage and a new outlook. The Caribé Hilton gave a tremendous lift to Operation Bootstrap, a plan fostered by Governor Muñoz Marín to jack the island out of its economic doldrums. It made a winter resort out of San Juan, began a new industry for Puerto Rico, encouraged Hilton International to begin what later became a worldwide network. As a hotel plant, the Caribé Hilton was the prototype for a new style in hotel building that spread all over the Caribbean and around the world. The emerging nations, encouraged by the success in Puerto Rico and looking toward tourism as a new industry, would henceforth not be satisfied with the plans of any architect who merely wished to build in the local idiom. Status required the hotel to be tall, terraced, air-conditioned and equipped with a large pool, a coffee shop, several bars that served fancy drinks, and a nightclub, preferably with gambling. The Caribé also proved that the prior presence of a beach was not a prerequisite for a resort hotel. A beach could be manufactured. It also proved that a hotel did not have to be built in a place that had a long established reputation as a resort. A well-run attractive hotel in a land equipped with good weather and good transportation could create a resort.

The West Indies, which before the war had offered little more than a string of ports to visit on a cruise, were now becoming both a playland for island-hoppers who came by air and a destination resort to which the pale and the fatigued hobbled in winter to convalesce at length. The Caribbean was not only about to become popular, it was on the verge of being stylish. One who helped its social standing was a young, attractive Jamaican of English ancestry and airs whose previous servitude included a term as aide to the Duke of Windsor and a turn at running a fancy haberdashery in New York. He was John Pringle, son of Carmen Pringle, matriarch of a tony Jamaican inn called Sunset Lodge. After nine years in New York, Pringle, equipped with a small inheritance, put $3000 down on a plot of beach land outside of Montego Bay in Jamaica and, flying between Paris, London and New York, managed to accumulate twenty-six shareholders in a hotel and cottage colony. The shareholders agreed to build homes on the property, which, in their absence, could be rented to the frostbitten wealthy seeking sanctuary in the winter warmth. His resort, called Round Hill, gaily decorated by Guy Roop, opened on a dazzling night in 1953. Noel Coward was at the piano and the assemblage glittered with gilded names from the assorted worlds of high fashion, high politics and high society.

With Pringle's handsome wife Liz available as a resident cover girl and model, and the house always filled with theatrical and social names, Round Hill became the subject of articles and often of cover pictures and color spreads in *Life, Look, Holiday, Town and Country* and *Saturday Review*. It drew the young senator Jack Kennedy. The William Paleys had a house there. Senator Jack Javits rented one. The piano in the bar might be occupied in the evenings by Richard Rodgers, Arthur Schwartz, Cole Porter or Noel himself visiting from his own villa. Italian landed gentry mingled with British royalty, with anyone from Greer Garson to the Samuel Newhouses, publishers of *Vogue* and a score of newspapers, tossed in for good seasonal measure.

If the winter life sparkled at Round Hill and the more peasanty confines of nearby Montego Bay, it was elegant in quite a different way on the remote atoll of St. John in the American Virgin Islands. Here Laurance Rockefeller, a fledgling hotelier, had bought the old Caneel Bay Plantation, which had been built as a company inn by the Danish West Indies Company. Rockefeller brought down New York decorators to spruce up the interiors, built more cottages on the assortment of beaches and opened a resort designed for subdued restoration. There were plenty of soft sand, acres of turquoise water, cool cottages in which to read or sleep, a passable dining room, but there was no golf, no tennis and no night life. Anybody who was in search of *that*, Laurance said, would be better off in Miami. Caneel Bay became popular, too, with a certain well-heeled intelligentsia for whom Round Hill was far too dressy and too flossy. At Caneel Bay Rockefeller proved that location meant nothing if you answered the public need. To get there a guest had to fly to Puerto Rico, change planes for St. Thomas, ride from Charlotte Amalie to Red Hook, a long and dusty run by taxi, then take a boat for St. Croix. Gradually Rockefeller bought up most of the real estate on St. John and then turned it over to the United States as a national park. Who would come there no one would explain, but there it was, preserved and safe from developers and other despoilers.

If one grew bored at Caneel Bay, there was always neighboring St. Thomas, which was gay, sporty, bohemian, a warm-sea edition of Provincetown or Sausalito, full of cute and tiny shops and bars often tended by oddball bipeds. To add to its precious guesthouses, tucked away so quaintly on its hilly streets, St. Thomas added a whopping oversized hotel. From the first day it proved an unwieldy behemoth crewed by an untrained native staff most of whom had never ridden in an elevator before, much less operated one.

The most colorful island of all was Haiti, full of drums and dancers, clouded mysteriously with voodoo, some of it genuine. One drank the five-star Barbancourt rum and bought brilliantly hued primitive paintings. Flashy hotels went up in Port au Prince, but the theater people and the writers, the magazine people and the photographers who worked on the high-fashion slicks, from Avedon to Gielgud to James Jones (who was married in its dining room), all stayed at a rustic establishment called the Grand Hotel Olofsson. It was neither grand nor Norwegian, but it was run with such ineffable charm by Laura and Roger Coster that it became the darling of the cognoscenti and the subject of a major article in the *Saturday Evening Post*.

The Caribbean and, to a lesser extent, the Bahamas were a near endless treasure trove of surprises, with new islands being unwrapped each year and the end of the supply never in sight. But the islands as yet attracted only easterners and midwesterners. The westerners went south to eat in the grand restaurants of Mexico City, the Jena and the 1,2,3, and to stay in the new del Prado Hotel. From there they took the plane over to Acapulco, which was being discovered all over again. It had a new hotel, airy and modern, called the Caleta, the boys were back diving from the Quebrada cliffs alongside

El Mirador, and there was a nightclub where you could dance on the sand. In the United States the desert flowered with sun-seekers, too. The Arizona Biltmore, in Phoenix, was still popular, but it was a holdover from another age. Across the cactus-strewn mesquite bloomed new ranches created not for cows or cowpokes but for dudes. At least one of them, the Casablanca, equipped with onion domes for decoration, also had separate closets for riding outfits lest the perfume of *eau de horse* permeate one's best duds. Wickenburg, not far away, became the self-styled dude-ranch capital of the world, and by dude ranch it meant a resort with horses, a swimming pool and all the help dressed up like cowboys.

California developed its desert, too, but it made no pretense at being a redoubt of the old woolly West. Palm Springs, a favorite of the stars before the war, now found itself all but overrun with popularity. Green golf courses sprang, one after another, out of the dry desert wastes. When the Thunderbird Club got stuffy over which stars with which racial backgrounds it would accept, the outs simply created their own club and called it Tamerisk. Some stars who weren't ruled out of Thunderbird joined both clubs, and real liberals, who weren't ineligible for Thunderbird either, joined Tamerisk. Although its large hotels, such as Mirador and the Desert Inn, had been the favorites before the war, Palm Springs now developed a wide assortment of small motel-like resorts which offered comfortable quarters, breakfast, a swimming pool and a place to get the sun. In Palm Springs, which never called a trailer anything but a mobile home, the new sanctuaries were never motels, always hotels, no matter their size.

Nevada had a desert, too, and out in the middle of the dullest flat acreage it had suddenly bestowed new life on an old Mormon settlement called Las Vegas. The Mormons had been attracted by the artesian wells that sprang out of the otherwise arid Spanish Trail. That's why the place had been named Las Vegas—the meadows. It had become a railroad town in 1905 and was incorporated by 1911, but it barely stirred until the Hoover Dam moved off the drawing boards in 1928. The curious could visit the dam, the playful could frolic on the waters of Lake Mead, the health-seekers could put up at Death Valley, and the athletes could ski at Mount Charleston. But that left Las Vegas with only fringe benefits until the state legalized gambling. From that simple decision a whole garish City of Chance grew in the middle of the barren desert. In downtown Las Vegas giant neon signs urged the passersby into the gaming parlors. Slot machines appeared in grocery stores and washrooms. McCarran Field, so dutifully named for the Nevada senator who had so diligently conspired to make sure that the laws permitted only the best people to visit our nation, was not only a glut of slot machines, it also had a coin-in-the-slot oxygen machine for those about to leave town with a hangover.

Las Vegas could offer a marriage service in a chapel at any hour of the day or night. It had once had an inn (and perhaps still does) called the Par-O-Dice Motel. The girls who danced in its shows were as bare as the Folies Bergère long before the topless vogue was legal or at least acceptable in San Francisco. It presented, primarily as a come-on, the starriest roster of entertainment names ever assembled anywhere. It has imported whole shows from Paris and from Tokyo. The hotels along its strip rivalled Miami Beach and all had freewheeling gambling all day and all night, no farther removed than the front lobby. One hotel that tried to make it without gambling collapsed in a quick heap and was forced to change its puritanical ways.

Only 288 miles from Los Angeles, Las Vegas took on importance not only as an uninhibited playland but also as a convention center—it built a huge hall for the purpose—and has become something of a national oddity. Like the Grand Canyon, it must be seen by people making their way across country. Airline rates, particularly from Los Angeles and San Francisco, dropped to new lows, but hotels and syndicates organized all sorts of lures including refunding air fare and giving the player a few dollars' worth of chips to get him started. Compared to the gentlemanly gaming rooms of Monte Carlo or even of Havana, not to mention the illegal ones that flourished for a time across the Hudson in New Jersey and in Palm Beach, Miami and Montauk, Long Island, Las Vegas's cigar-smoking, loud-mouthing, crap-shooting style had not been seen publicly since the raffish days of the early West.

By late 1953 American Airlines had placed its new DC-7s in cross-country service, cut the flying time between California and New York by a startling three hours. Nonstop flights were scheduled at 7 hours and 55 minutes westbound and 7 hours and 15 minutes eastbound. The first east-to-west flight was forced down by the Rocky Mountains and had to put into Denver. On the return flight, however, American was out to redeem itself. Flying five miles up and reaching speeds of better than 400 miles an hour, Captain Harry Clark brought the plane into Idlewild Airport in 6 hours and 31 minutes, cracking by a full quarter of an hour a record that had been held by a United Airlines' DC-6 since 1947. The record run of American's DC-7 had come only a few days short of that moment fifty years before when the Wrights flew for twelve seconds at Kitty Hawk. And it was just twenty-six years after the old Fort Trimotor that roared from New York to California in 54 hours including two full days in the air, eighteen stops, one night in a Texas hotel and an overnight run by rail.

Now it was possible to cross the nation within the time of a working day. American was eager to display the disparate cultures that flourished on either coast. For the inaugural run it brought a group of western writers to the East while the eastern group, many of them big-name writers whose columns were syndicated all over the country, were exposed to the West. The eastern columnists were taken to a football game between UCLA and USC where they watched perplexed while the cheerleaders called for such rhyming cheers as "Hit 'em with iron, Myron," and "Pack the ball, Paul." They were even more wondering when at half time the cheering section broke into their placard-flipping act creating pictures for the opposite stands. The scenes included tableaus in favor of stopping forest fires and exhortations to buy savings bonds and to give blood. That night American arranged a typical Hollywood party in a house where the spigots poured whiskey, a waterfall could be operated by remote control from the owner's car, a tree grew through the bedroom, and the swimming pool was in the living room. Seven people fell in during the party. A somewhat shaken easterner, thinking back on the strange tribal customs performed at the football game and the oddities of the Hollywood party that followed, cleared his throat and said, "Do you realize that with this new DC-7 nonstop, coast-to-coast service these people will be less than eight hours from our front door?"

The Far West was, however, a travel attraction in itself. One went to Hollywood to see the stars. Disneyland was to come a bit later. In Beverly Hills and in the canyons one could still find the street sellers hawking maps of the stars' homes. Hollywood was the only place in the world where the sideshows were a drugstore and a cemetery. One went to Schwab's to see how the stars lived as one went to Forest Lawn to see how they died. Forest Lawn was one of the few cemeteries around that employed a press agent. But then it also invited tourists, who could see the works of art, take in the mausoleums and the plots where the great men and women of the screen repose, not forgetting to stop for souvenirs on the way out.

To the north San Francisco was everybody's delight. No one liked it better than the Europeans, who, despite the restrictions, began to trickle into the country on business missions. One rode the cable cars, toured the harbor, peered at Alcatraz, climbed up the hills to Julius's Castle for the pot roast, rolled down the hills for the steamed crabs on Fisherman's Wharf. It was newly celebrated as the birthplace of the United Nations, but still renowned as the gateway to all the mysterious lands that lay along the Pacific basin.

The Far East rebounded more slowly than Europe from the effects of war. For one thing, relatively few Americans had been pleasure travelers in the Orient before the war. The distances were great, the travel time was long, the expense was formidable and the scene was still troubled. With its first postwar ships not yet ready, and the demand for travel to the east running heavy, American President Lines pressed the *General Gordon* and the *General Meigs*, two unconverted troopships, into passenger service. Even after the *President Wilson*, its first new ship, appeared in December 1947, and the *President Cleveland* entered service in the spring of 1948, the *Meigs* and the *Gordon* continued to sail. The *Gordon* was diverted to Tientsin and Shanghai to evacuate the Western capitalists when the Communists took over China in May of 1949.

In the early postwar days, Japan remained under military occupation so many businesses preferred to operate out of Hong Kong. The Peninsula Hotel on the Kowloon side was so crowded that Arthur Chase, an agent for American President Lines who lived there for six months, found himself almost every night with a new roommate installed by the management. Jimmy's Kitchen, the Parisian Grill (popularly called the PG) and the Grips in the old Hong Kong Hotel were the favored places of congregation. Only five stories high and delightfully Victorian, the Hong Kong Hotel was doomed in the boom. Its site, on Peddars Street and Queens Road, next to the Gloucester Hotel, was too valuable and it was knocked down in favor of a large office building. The Gloucester Hotel became an office building, too, and hotel and tourist life reverted to the Kowloon side, where it was to stay until midway in the 1960s.

Tokyo tourists, even into the early 1950s, were consigned to the Imperial Hotel, which had been designed by Frank Lloyd Wright. It was a marvel because it had withstood the earthquake of 1923 (which killed 74,000 people and demolished 64 percent of the city's buildings) and even the American bombings of World War II. But the great building and all its facilities were scaled to a diminutive Japanese race. Westerners had to duck under doorways and bend over in a deep bow to use the washbasins. Food outside the hotel was sketchy at best, and only the old Oriental hands tried it. A modern version of a hotel, the Nikkatsu, opened in Tokyo using several floors of a new office building. The trains operated with dispatch and with style. Aboard, the *Swallow*, a crack train that sped from Tokyo to Kyoto, the club car attendant wore an armband with the word "Boy" written on it. All porters, conductors and candy butchers stopped, bowed and lifted their caps when the *Swallow* flew out of the station. Midway in the journey, the train paused and all passengers were invited out to the platform to participate in setting-up exercises conducted by a physical training instructor.

The *Wilson* and the *Cleveland* opened the first luxury travel to the Far East carrying 350 passengers in first class and just under 400 in steerage. The first-class passengers were missionaries, businessmen, wives and dependents en route to join husbands and fathers in occupation service in Japan, and some Orientals on official missions. The steerage was a dormitory divided by sexes and usually reserved for Oriental passengers only. APL, like most Americans, still thought in colonial terms. They were sure no westerner would put up with it.

Barnett Laschever, who was later to become travel editor of the New York *Herald Tribune* and tourist director of Connecticut, was one who did. Caught yenless in Japan, he made the trip in 1951 with his wife. She was assigned to a dormitory with eight women—Japanese, Chinese and Filipino —he to a room with four Chinese men. The high point of the evening was the ten o'clock noodle party. "We all lined up with our little bowls and were served a bowl of noodle soup," he recalls. "The food was imitation Chinese thrown together by inept American cooks. Up top the white folks were having a costume ball, vying with each other in their Oriental costumes. Down below, in steerage, the Orientals were trying their best to look like westerners."

The missionaries often busied themselves learning Cantonese, which was spoken in Hong Kong. Many already knew Mandarin, but that Chinese language was useless now that the Communists had taken over the rest of the mainland. The matrons in their trig suits, with husbands deceased and nothing to do at home, were already on the high seas on their way around the world. Army brats ran the gangways like unpenned ponies, but for businessmen it was all more like a cruise. The Royal Hawaiian Band was out to serenade the ships as they edged into Honolulu harbor, and there was a day to swim and surf in Waikiki before the ship moved on. The tables bloomed with the fruits and flowers of the islands after the Hawaii stop and the kitchen sent out broiled mahi mahi and pineapple fritters frothy in a foam sauce. Japanese passengers often called for the fixings and made sukiyaki right at their tables. The run took nine days from Honolulu to Yokohama, and another two days to Hong Kong. It was a slow boat to China, but for all except the steerage-class passengers it was one of the fascinating journeys of the time.

Meanwhile, back on the Atlantic, the tourist migrations were already in full seasonal flower. On July 3, 1952, America made its own major bid on the North Atlantic run with the *United States*, a speedy 990-foot superliner. She was not as long as the *Queen Elizabeth*, nor was she as fancy as many ships. She had been built under strict naval supervision, and for a nation still skittish from the last war, it was pointed out that she could take either 2000 passengers to pleasure lands or 14,000 soldiers to combat. While defense specifications, as the naval architect assured the press, came first, some frippery was allowed. She was, for instance, the first ship to make use of that new decorator's color combination, blue and green. A cocktail lounge was brightened with simulations of Navajo sand paintings, but on the whole she was, when she first arrived from Newport News, a quiet, refined, almost stuffy ship. She did have speed, however, and when she won the blue ribbon, it was the first time an American ship had held the Atlantic prize since the S.S. *Baltic* of the Collins Line sailed out of Liverpool and made it to New York in 9 days and 13 hours. That was in 1852.

The flashiest ship ever to float sailed into New York Harbor in 1953. She was the *Andrea Doria*, named after a famous Italian admiral, whose bronze statue reposed in the first-class lounge. To one writer she resembled a "dream sequence from one of Hollywood's mammoth musicals." In some staterooms, designs of blue fish ran all through the white plumbing. The unofficially named Rita Hayworth Suite was cloaked in quilted satin walls, hung with satin draperies and covered with satin bedspreads. One suite had white leather walls, except for a single panel which was entirely covered by a tapestry. The *Andrea Doria* had three outdoor swimming pools, a testimonial to the weather on the Mediterranean run. She carried a children's dining room complete with flower-munching pelicans and smoking frogs on the walls. She had four movie theaters, a winter garden with giant mosaics on the walls. When she went into eight-day express service between Naples and New York after a shakedown West Indies cruise, she was sold out, including the Zodiac Suite, where the fare for two was $1000. Each.

On a July night three years later she was to collide with the Stockholm and sink in the sea off Nantucket Island. The incredible accident and the sinking reached into the most hardened hearts. Said a New York cabdriver, "It was like losin' a woman." Said Harry Manning, who had been skipper of the *United States* when she won the blue ribbon of the Atlantic, "The art alone is enough to make a man weep at the loss of it." The stories went on for day after summer day. The New York *Times* thought it would have an eyewitness account. After all, Camille Cianferra, an ace foreign correspondent, was on board. But Cianferra's cabin had been directly in the path of the oncoming bow of the Stockholm. He was killed outright and his wife's daughter, the daughter also of broadcaster Edward P. Morgan, was missing. She was found a day later aboard the Stockholm, which had somehow carried away her unconscious form when it withdrew from the Doria's gash. She was still alive. All that sorrowful fall the lawyers and their witnesses played out the sad charade in the courtroom on Foley Square where many of America's most famous trials took place.

The loss of the *Andrea Doria*, as deeply felt as it was, didn't deter the rush to Europe. Trade-not-aid was the catchword of the day, especially in Washington's official circles. Representative Jacob J. Javits, then in the House, sponsored a bill in 1954 to establish a United States Travel Commission. The commission was not to encourage travelers to come to America, as a later bill did, but rather to send as many Americans as possible on foreign pleasure tours so we could help the free nations. "Helping foreign economy by sending Americans on foreign holidays," Javits said, "offers the taxpayer a program in which he can participate with the greatest possible pleasure." Noting that there were already 12 million Americans earning $5000 a year or more, Javits called for reduced fares that would permit the broad middle class to take a two week's trip for $500, transportation included. He ridiculed the papers for getting "all excited because some factory sent a hundred people abroad in two planes." They called it, he chided, the largest mass commercial flight in history. "Why, we have 60 million people gainfully employed," he said, "16 million of them in factories." He thought them all a likely market, but the bill never got anywhere.

Fares, at least in the air, were coming down, though admittedly not enough to reach the broad-middle-class traveler who chafed to go. Pan American began its Rainbow Service to Europe in time for the summer season of 1952, bringing two classes to transatlantic air travel. The other airlines followed. In 1954 tourist rates were introduced on the Pacific, bringing the new second-class rate from San Francisco to Tokyo down to $488, more than $160 cheaper than the first-class rate. Hardy trippers could now fly around the world for $1100, an experience hitherto possible only for someone prepared to pay. By May of that year, Pan American announced that it was now prepared to sell tours anywhere in the world, or around it, for that matter, on the installment plan. A writer assessing the new idea wrote, "We understand how a piano can be collected for nonpayment of installments, but we were at a loss to determine how a finance company goes about retrieving three weeks on the French Riviera. Also, you can alway change your mind about the piano and send the ruddy thing back. But suppose your two weeks in sunny Bechuanaland resulted in two weeks of rain, a bout of scurvy and an invasion of tsetse flies. How do you feel about paying for it over the next two years?"

Pan American wasn't worried. It noted that the automobile industry, if not the whole American economy, was based on buying on time. "By the time you repossess a washing machine it's obsolete anyway," said an airline man, who added, "Besides, 60 percent of all Cadillacs are bought on credit."

The loans, which could be processed in twenty-four hours, might well include hotels and sight-seeing as well as transportation overseas. Go now and pay later became the new byword and people leaped at it. Nor when the tensions increased in Europe was there any slackening of reservations. When ominous clouds gathered over Berlin, it was smart to say, "I'm going to Europe. I want to see it before it closes."

France was popular, but Italy was the first real find.

America took Italian Fascism as something of a joke, and no one was stopped from visiting Italy merely because it had sided with Germany and Japan. The Excelsior in Rome became an American outpost, and the outdoor tables at Doney's on the Via Veneto became an American sidewalk café. Americans trooped through the Pitti Palace in Florence, soaked up the sun at Lido Venice, crammed into Harry's Bar for hot hors d'oeuvre, climbed the narrow roads of Capri, admired the yellow-tinted columns at Agrigento on the coast of Sicily, discovered Positano outside of Naples, gasped with mock shock and secret delight over the pornographic statues of Pompeii, sailed again on the quietude of Lake Como. If anyone made Italy for Americans it was a doughty, robust Italo-American lady named Manolita Doelger, who was the unfailing tourist commissioner for Italy in the United States during the years of the buildup.

Spain was quite another story. With its Franco fascism, it had stood behind the Axis cause. This kind of fascism Americans took seriously. What's more, it was still the law of the land. Politically sensitive travelers thought twice before supporting Franco, but there were people who either didn't have political compunctions or were just plain curious. Returning from Spain, they told of the small inns called *paradors* dotted around the countryside. They told of the prices, which were frequently a fifth the cost of trips elsewhere in Europe. If there were trappings of a dictatorship, they were at least not in plain view. The country for all its fascism was attractive and even exciting. There were the comforting thoughts that Hilton had opened a hotel in Madrid. That Hemingway was back. And finally the incontrovertible fact that the American government was giving aid to Spain and building bases there.

In the great rush that developed during the last half of the Fifties, American travelers were everywhere. They discovered Ireland and went pub crawling along the Liffey. They invaded England in irrepressible hordes, and hotel rooms in London, overpriced as they were, were always at a premium. Everybody came home complaining about English cooking and some even about English weather, but it did nothing to diminish the popularity. Visit Britain first, the veterans said,

and you won't be disappointed. A traveler who took in a chop
suey palace in Soho came away shaking his head. "Even the
English Chinese can't cook," he said.

Denmark became an object of considerable affection, and
the film *Hans Christian Andersen* and the lilting song called
"Wonderful Copenhagen" did little to diminish its sudden
popularity. Travelers sailed the fjords of Norway, and jockeyed
for reservations aboard Sweden's famous *Dollar Train*, which
took excursionists in great splendor into the Land of the
Midnight Sun. There they met Lapps, played golf at midnight,
drank akvavit and danced in Dalecarlia, read American papers
every day, swam in brooks while the train waited at a siding,
took hot baths, had their clothes pressed, and went on endless
excursions. Pioneers went on after that to Finland, where the
imaginative ceramic work was a revelation and so was getting
beaten with a birch branch by a hefty lady attendant in a
sauna.

In 1955 Hilton flung a hotel in faraway Istanbul and Americans flocked to it. Elsa Maxwell and the assembled royalty
of Europe had cruised the Greek Islands, and there was a rush
to charter a yacht, book a seat on the aging Semiramis, or
buy an airplane ticket and make tracks for Rhodes, Mykonos,
Delos, Santorini and other whitewashed islands with strange-
sounding names. In 1955 a planeload of men, tended by one
lone blond stewardess, flew off from Honolulu Airport in a
chartered DC-4 that groped its way through the night, refueled on Canton Island and then touched down the next
morning on the storied atoll of Bora-Bora in French Polynesia.
A great flying boat up from Australia waited for them in the
harbor and soon it thundered over the water and flew them in
an hour's time to the dreamland of Tahiti. The passengers had
paid $1000 each to be the members of the first air tour of
Tahiti. The hotel accommodations left quite a bit to be desired,
but the island didn't. It fulfilled the expectations of all those
who made that trip, except perhaps for the stewardess, who
had been queen of the flight, but that was before the first
Tahitian beauties wiggled into sight.

The Middle East seethed with animosities, but that didn't
seem to hold back American tourists. One could be caught

in Baghdad by rioting mobs, as some were. One could be arrested in Damascus for carrying a tape recorder in the marketplace, as one was. One could be turned back at the border of an Arab country if one's passport was stamped with an Israeli visa, as many were. One American dress designer from New York's wholesale garment district on Seventh Avenue was aboard a plane that made a stop in an Arab country. Occupants were required to fill out a questionnaire before they were even allowed to wait in the airport while the plane was being refueled. Jews were required to stay aboard the plane. When it came to the question of religion, the designer thought a minute then wrote, "Seventh Avenue Adventist." She was waved inside. Crossing from Jerusalem in Jordan to the other part of the city, which lay in Israel, was a tricky business. It required a Jordanian visa and an Israeli visa, but the Israeli form had to be on a separate piece of paper or the Jordanians would not recognize the passport. In addition, a military permit was needed from both sides. The Israelis needed three days to put one through, not including from sundown Friday through sundown Saturday. The Jordanians could get theirs done in two days, not including Friday. Most permits had to be obtained through the American consul and he didn't work on Sunday, so that anybody arriving in Jordanian Jerusalem on a Thursday and desirous of crossing into Israel could count on spending a long weekend.

The tourists crossed at the Mandelbaum Gate, barred on the Arab side with a cement barricade slit with peepholes, and guarded by Arab legionnaires with red scarves on their heads and unsheathed bayonets on their rifles. Concertinas of barbed wire were strung among the cement tank traps. American Express couriers struggled with their herds, wiggling them through the formalities, but independent travelers did it on their own, crossing the eerie no-man's-land with a Jordanian porter carrying the bag to the halfway point. There he dropped it and an Israeli came to pick it up. In the still ruins stood an empty house with a sign lettered in English and Hebrew. "Welcome Within Thy Gates, O Jerusalem," it read.

In the Soviet Union, Khrushchev had assumed total power and slowly the Curtain began to part. An American company of *Porgy and Bess* went to Russia accompanied by such a mixed bag of camp followers as Leonard Lyons, Truman Capote, Ira and Lee Gershwin and Harold Arlen. The first American travel writer to enter Russia since the sinister purges of the late 1930s was granted a visa. In the late winter of 1955 he flew from Helsinki to Leningrad in a Russian plane that neither warmed up for the takeoff, nor provided a hostess, nor carried a seat belt. For the customs inspector he filled out a form avowing that he carried no precious metals, precious stones, pearls, platinum, gold or silver, no Soviet currency, U.S.S.R. State Bank Loans, no promissory notes, no ammunition, field glasses of sixfold power or higher, no wormwood heads or seeds, opium or hashish or pipes for their use, no horns of steppe antelope, or Manchurian deer, unstiffened horn or maral or spotted deer. All meals, hotel rooms, transfer services, sight-seeing guides and private sight-seeing cars had to be paid for in New York. In that way he toured Leningrad and Moscow, flew south to Yalta, saw the Crimea, ventured along the Black Sea Riviera, poked into Georgia, covering some 2000 miles, every step of the way with a Soviet-assigned interpreter. It was a hard trip and a trying one, continually grinding against an alien and hostile national concept, but never for one waking instant was it less than fascinating.

An account of a train trip from Leningrad to Moscow experienced by a Western tourist in the late winter of 1955 was an accurate picture:

> The midnight express that rolls the 400 miles from Leningrad to Moscow every night of the week is called the *Red Arrow*. For the nine-hour excursion from Russia's window on Europe to the present center of the Communist world foreign tourists are offered space in the "soft" car or the International Car. I'm not at all sure that you'll have a better chance to meet and mingle with Russians in the "soft" accommodations than in the International Car, for the night I rode the *Red Arrow* I didn't come across another foreigner on the entire train. A ticket in "soft" will net you one of two lower berths in a private compartment, with the bath down the hall. When it

comes to roommates you'll have to take potluck.

On the International Car, a class I came to think of as "softer-than-soft," you draw an upper or a lower berth in a private compartment that also contains an uphol-stered chair, a permanent table and lamp, blue plush drapes (which the Russians call "ploosh"), café curtains, canned music, and a bathroom that connects with an ad-joining compartment. It costs $20 extra for the Inter-national Car but, although there's no telling who will be in the other berth here either, chances are you'll meet a better class roommate on the softer-than-soft.

Business was slow my night on the *Red Arrow*, for I drew no roommate at all. My shareholder in the joint bathroom was a Russian general. A samovar boiled gently through the night, watched over by a white-coated at-tendant. We made one stop. In the small hours of the Russian night I was sent scattering by the sudden blare of a loudspeaker in the depot. It was 5:20 A.M.

Some three hours later, when I finally arose, I was just about to enter the joint washroom to shave when I saw the indicator on my side flip to a word that un-mistakably meant "occupied." While I awaited the gen-eral a lady vendor knocked on the door, there being no diner, and offered cheese, salmon, or caviar sandwiches. Well, one *had* to choose the caviar. It might be recherché, but it was one way of having eggs for breakfast.

Any Westerner en route to Russia in those days was sure to get a long string of objections from family and friends. One West German businessman, staying at Moscow's National Hotel, confided that he had to cable home every night. An American who was trying to work out an itinerary with a representative from Intourist, the Soviet Government travel agency, expressed some concern about flying out of Russia on the last day his Soviet visa was valid. "Suppose the plane is delayed, and I'm required to stay another day?" he inquired not without some nervousness. "Then," boomed the Russian travel man as he puffed on a cigarette held between thumb and forefinger, "then we wicked Bolsheviks will send you to

Siberia." The trouble was, said the traveler who later related the story, you couldn't be 100 percent sure he was kidding. The Intourist interpreters would display visible pique if tourists photographed too many soldiers or shabbily dressed people. And a picture of a rail station or a river with a bridge over it could bring arrest. Hotel rooms were said to be bugged and visitors were said to be shadowed if they left the hotel without their guide.

Russia, with its sense of adventure, its aura of the forbidden, of possible danger, became, that summer of 1956, and for many summers to follow, the trip to take. In the sophisticated parlors of America one could hardly be an engaging guest at a dinner party if one wasn't recently home with tales of explorations in darkest Russia.

Travelers of sophistication were digging into every cranny of the world, seeking new and more unusual places that the ordinary traveler, much less the tour groups, were yet to discover. Douglas Aircraft's DC-7, last of the piston planes, was refined, enlarged, given greater range. So was Lockheed's Constellation, called the Super Starliner. Air France sent one from New York clear to Athens nonstop, making the crossing in 14 hours. Still the ships came and went, disgorging huge numbers of travelers in England, France, Italy, Sweden and Holland. There were days when five mammoth liners were tied up in a row against the piers on Manhattan's Hudson River. In 1957 the ships carried a record 1,032,400 passengers to Europe. In September they would all come home in a rush creating long waits for customs inspection and incredible tangles of taxis, porters and baggage at New York's decrepit, obsolete piers. The ocean liners were riding the crest of the waves, shrugging off the storm warnings that rumbled out of the great aircraft plants along the Pacific coast. Then in 1958, the noise sounded shrill and clear and overpowering, a ringing that would change travel habits and patterns, cause revolutions in hotel planning, alter the character of resorts, create new business practices, forge a whole new social caste, lift entire nations out of remote, unreachable obscurity. It was a high whine, and you could scarcely listen to it with the naked ear. It had authority and it couldn't be dismissed. It was the sound of the jets.

IX | *The Tourist Diaspora I (1958–1969)*

THE CHINESE CALLED IT the Year of the Dog, and the Orthodox Jews numbered it year 5716. It was Anno Domini 1958, memorable as the year in which planes began flying without propellers. It was the start of an era of moment, a time when the airport runways would be lengthened and the world would shrink so that no place was more than twenty-four hours from any other. The new American-made jets would carry nearly twice the number of passengers as the old DC-7s. On the westbound flight from Europe they would arrive in the New World, considering the time change, only a few hours after they left the Old. Surely there would be revolutionary changes in hotel planning, among many other facets of travel, but of all the world's travel planners, the French were displaying the most sangfroid. A delegation of Paris hoteliers, where a new hotel had not been built since the Georges V went up in 1933, arrived in New York just as the jet age was about to unfold. Would they at last build new hotels in Paris, the press asked? Ah, non, they said with exquisite Gallic logic. It stood to reason, didn't it, that if the customers from America would get to France twice as fast they would also leave twice as fast, so therefore no new rooms would be needed. Q.E.D. Then they flew back to France on a piston-engine plane totally assured their theorem was right.

The jets were due in the fall, but before they rose in the skies, there were other shaking events. The airlines, looking forward to the big planes and envisioning the rows of empty seats that they would have to fill, agreed upon a third class of transatlantic service. It would not be called steerage, as the early ships had termed it, but economy. The new economy class would forswear the elaborate meals, the berths, the stretch-out seats, the footrests, the free liquor and the blizzard of gifts. In economy class the airline agreed to serve no meals except for a sandwich. Just what constitutes a sandwich touched off in the spring of 1958 a memorable conflict known as the Great Sandwich War.

French correspondents in the United States filed stories to their newspapers in which they called the fracas "La Bataille du Sandwiches." From their positions among the forward mustard pots, correspondents stated the case of the Americans versus the Europeans. Pan American and TWA were very explicit. They would serve such sandwiches as ham and cheese. TWA explicitly issued instructions that its sandwiches were to measure no more than 4½ inches by 3½ inches by ½ inch thick. Directives went down to the commissary showing the exact location of the allowable embellishments: a whole gherkin or a radish rose.

But not so the Scandinavians, who came from the land of smorgasbord. Oskar Davidsen's in Copenhagen, they readily recalled, served 177 different kinds of sandwiches. SAS served liver pâté, fried bacon, mushrooms, sliced tomatoes or seven slices of ox tongue, slipped a piece of bread underneath and called it a sandwich. The French got wind of this and cried foul. They began to make *pâté de foie gras* sandwiches with caviar on top and threatened to serve a whole *coq au vin* with a piece of bread underneath it. Voilà! A *coq au vin* sandwich. At this, Pan American registered a complaint with the International Air Transport Association and the battle was on. On television across the nation that night, Chet Huntley flipped the ball to David Brinkley and David said, gravely,

Tonight an international crisis about sandwiches . . ."

Reported Lowell Thomas on CBS:

A new international question brings more trouble in world affairs. When is a sandwich not a sandwich?

From London, Alexander Kendrick of CBS reported to the United States:

The ninth Earl of Sandwich is eighty-three years old and lives in seclusion with his books and memoirs. Whether his illustrious ancestor, the fourth Earl of Sandwich, rests so quietly, however, is less certain, for there has burst upon the air the greatest controversy about sandwiches since the fourth Earl, back in about 1750, clapped two slices of bread to a slice of ham and made himself a name in history.

Finally the august judges of the airline association met and handed down a ruling. Immediately the cables crackled. Here is the wire that went out from Pan Am offices on the West Coast:

LAXDH PDXDH SEADH HNLDH CPY IDLXZ MILLER FOLLOW-ING FOR IMMEDIATE RELEASE AT YOUR EDITORIAL JUDG-MENT BEING HANDED WIRES HERE COLON

PAN AMERICAN AIRWAYS TODAY REVEALED ITS VERSION OF THE DAGWOOD/THE SUBSTANCE AND CONTENT OF SANDWICHES SERVED ON ECONOMY CLASS FLIGHTS OVER THE POLAR ROUTE BETWEEN THE WEST COAST AND EU-ROPE.

RULINGS OF I.A.T.A. COMMA THE INTERNATIONAL AIR-LINE AUTHORITY COMMA BARRED CAVIAR COMMA PATE DE FOIE GRAS COMMA OYSTERS AND OTHER DE LUXE INGRE-DIENTS. SANDWICHES SHOULD BE COLD COMMA INEXPEN-SIVE AND A SUBSTANTIAL PART SHOULD CONSIST OF/BREAD COMMA ROLLS COMMA OR BREADLIKE MATERIAL/WAS THE DICTUM HANDED DOWN.

SANDWICHES CAN BE OPEN/FACED OR CLOSED. ESCHEW-ING PEARLY GRAINS OF CAVIAR OR OTHER HIGH/TONED DELICACIES COMMA PAN AM OFFERS HONEST AMERICAN SANDWICH FILLINGS LIKE ROAST BEEF COMMA ROAST TURKEY COMMA BAKED HAM AND EVEN SCRAMBLED HAM/AND/EGGS FOR BREAKFAST SANDWICHES. FRENCH/CROIS-SANT/ROLLS AND BUTTERED COFFEE CAKE ARE ALSO BREAKFAST ITEMS.

THE LUNCH OFFERING COMMA FOR INSTANCE COMMA IS A TRAY HEAPED WITH BREAD TOPPED BY SAN FRANCISCO BAY SHRIMPS WITH LETTUCE COMMA TOMATOES AND SLICED EGG COMMA FLANKED BY HAM AND CHEESE AND

TURKEY SANDWICHES COMMA GARNISHED WITH CRAN-
BERRY SAUCE COMMA LEMON SLICES COMMA PARSLEY
COMMA PICKLES AND OLIVES.

DINNER SANDWICHES FEATURE OTHER VARIATIONS ON
OLD/FASHIONED ROAST BEEF AND HAM. SECONDS ARE
FREELY AVAILABLE.

UNDER I.A.T.A. RULES COMMA COFFEE COMMA TEA
COMMA MILK COMMA AND MINERAL WATER ARE SERVED
COMMA BUT ALCOHOLIC BEVERAGES ARE NOT. SFODP WILEY
151800

When the Great Sandwich War finally subsided the nations
of the world could turn to more pressing problems—the fair at
Brussels, for one thing. The gates opened on April 17, 1958,
nineteen years after the last world exposition in Flushing, New
York, a memorable fair that perished with the onset of war.
Now the nations were on postwar view, their rivalries, alle-
giance and the power balances somewhat altered. National
personalities were, however, the same. The daring cantilevered
building put up by the French had sunk in the spring mud;
on opening day the French pavilion was a mass of packing
crates and confusion. The Italians were late, too. The Swiss
opened exactly on time. The Germans had finished their building the prior fall, covered it with a tarpaulin and gone home
to await the spring. But the spotlight was on the Americans
and the Russians, for their political rivalry invaded the fair-
grounds and the fair's great interest was the spice of competition between the two nations.

The United States had filled its round pavilion, designed by
Edward Durell Stone, with vestiges from American life: voting
machines, a giant Idaho baking potato, paperback books,
restaurant menus. It was a static show except for an IBM
machine that memorized historical facts from 4 B.C. to the
present day, an electromechanical manipulator for handling
radioactive objects, and *Vogue*'s live fashion show. What saved
the American exhibit was Walt Disney's Circarama, a stand-up
theater that portrayed glimpses of America and its people on a
360-degree full-circle screen.

Spending an estimated $60 million on their exhibit, the
Russians built a big glass rectangle reachable by a long prome-

nade up a monumental stairway. A pair of massive statues of
workers at either side of the portals were dwarfed only by the
image of Lenin at the far end of the hall. A border of heavy
machinery lined the way to his feet. But near the entrance of
the building the Russians had placed their strong suit, a
sputnik, silvery and shining, its beep-beep sounding an inces-
sant call over the loudspeaker. Sputnik had only recently been
lofted into the stratosphere, giving the Russians a frightening
lead in the space race.

New York's Brass Rail restaurants ran the food concession
at the American pavilion, and while many visiting Americans
complained about the high prices and the cold pastrami, the
Russians were having their own problems. A party of five fair-
goers reported that dinner in the Russian pavilion had cost
them $62. Six lunch guests came away with a bill for $82 and
three who stopped for vodka and caviar, the first course in an
international progressive dinner they had planned, paid $12.

The word soon got around that the best food at the best
values was available at the Czech and Hungarian pavilions.
Both exhibits created great interest anyway, for the countries
were making their initial public appearance under Soviet
domination. The Czechs brought rivers of Pilsner beer, served
outdoor lunches for $1.50. The Hungarian Restaurant quickly
garnered the title as the best of the fair, and it was never
challenged. The Soviets opened an ice-cream parlor on the roof
of its movie theater, with seats for 650. The Americans
countered by bringing in a popcorn machine, but some profi-
teering Belgian had already registered the word "popcorn"
with the Fair authorities and demanded a royalty for every
kernel popped. The American corn poppers withdrew.

The Russians scored heavily by bringing in the Bolshoi
Ballet with its great star Ulanova, then in her last performing
years, followed her with the Moscow Circus, the Moiseyev
Ballet, the Obrasov Puppets, all of which were put on in
August, when no other major cultural events were scheduled.
The Americans countered with Benny Goodman and Harry
Belafonte and a touring company playing Rodgers and Ham-
merstein's *Carousel*. Leontyne Price, Blanche Thebom and
Isaac Stern with the Philadelphia Orchestra were our longhair

entries. But there was bitterness that Hollywood, whose internationally known stars had been asked to appear even for a walk-on, all proved too busy to be in Brussels during the hot cold war of 1958. The hand-wringers at home worried that having lost the first round in the space race, America was now losing the cultural race, too.

Brussels was the prime attraction that summer of 1958, but travelers were also down in Spain sunning themselves along the Costa Brava, where new hotels were rising alongside every sandy cove. They were up in Berlin to see the cold war firsthand, flying that narrow corridor into the city and landing at old Tempelhof Airport. The Berlin Hotel was just being built, but there was the Kempinski, not exactly luxurious but still passably comfortable. The food was excellent in Berlin and in the warm weather one could order a tank of beer in shades of strawberry red and lime green.

On the high seas the *Cristoforo Colombo,* sister ship to the late *Andrea Doria,* was taking tourist-class passengers to any Mediterranean port for $200. The baths were down the hall, but what was once the steerage now sported an open-air swimming pool, tables for six in the dining salon and a dance band at night. Movies shown in first class were beamed all over the ship by closed-circuit television, the first system ever to be installed on a ship. There seemed no limit to how much some passengers would spend for the crossing. Some first-class apartments cost $1700 double, or $1325 if occupied by one person. The passenger lists might include Rossano Brazzi, who had appeared in the Ezio Pinza role in Hollywood's version of *South Pacific,* or six of Prince Faisal's sons on their way home to Saudi Arabia, or Bob Ruark, who then kept a home in Palamos on the Costa Brava. There might be a rich American matron sailing to Europe with her children for a summer on the Riviera, or a wealthy Egyptian telling fellow passengers,

> There is nothing quite like the life in Cairo. I have five servants. I pay them a ridiculous figure. They work seven days a week and 366 days a year on leap year and they love me. Love me! I ride each morning, I go to the club for lunch, I drop by the office in the afternoon. It is very

handily located. The other night I gave a party. I hired
an orchestra. Eight of them. They played all night. And
what do you think I paid them? Forty-five dollars. It's
nothing. Life in Cairo is fabulous and they say it's good
for another twenty-five years.

Those who flew that summer could take advantage of the
new economy-class service, but they could also relax in the
comfort of the Boeing Stratocruisers and sleep their way across
the ocean in a roomy berth. That kind of comfort would soon
be victim to the high-speed planes, although nobody thought
about it then. Another victim would be Shannon Airport,
where in that last year of transatlantic propeller planes 300
flights a week glided in for a landing. Since 1947 it had been
famous as a duty-free port, the first in the world to sell untaxed
liquor to transiting passengers. With nearly $60,000 in sales in
the first six months after it opened, it quickly added French
perfume, then German cameras, Waterford glass, German toys
and Scottish cashmeres. It had been the home of Irish coffee,
invented at Foynes Seaplane Base during the war. The local
coffee was so bad that somebody doctored it with Irish whiskey; only later was it embellished with whipped cream. There
was nothing that would restore the soul after a long Atlantic
flight better than an Irish coffee, a tonic that helped to popular-
ize Irish whiskey as well as the Irish. All that lay under threat
of ruin now, for Shannon, which was selling well over a quarter
of a million dollars in cameras in a year, would surely be over-
flown by the new jets. Its promoters were trying to save the
day by enticing light industry. Anything to keep Ireland green.

With the summer over and most vacation travelers back
home, the jet race got down to the wire. Britain was betting on
its Comet, a revised version of its 1952 jet that had very nearly
scooped the worldwide field by a neat six years. It had gone
into service between London and such distant points as South
Africa, the Far East and Ceylon. In 1954 the first of them exploded in midair, dropping thirty-five passengers into the sea
over Elba. Three months later another one blew up, and the
British launched an investigation. Three Comets in good condition were turned over for testing. Two were flown and the

third was subjected to severe pressure changes by submerging it in water and flexing its wings artificially every few minutes. After 9000 hours of this simulated flight, the metal skin of the plane tore apart, and the investigators had their answer. The cause of the crashes was metal fatigue.

The new Comet took four more years to perfect and now the race with the new American Boeing came down to a matter of days. The British won it on October 5, when they sent a Comet 4 racing from London to New York and another from New York to London. The eastbound flight broke the record by 4 minutes, requiring just 6 hours and 12 minutes for the crossing. They had not intended to make it a race, spokesmen for BOAC said, but then Pan American had run an advertisement saying it would inaugurate jet service over the Atlantic on October 27. That got the British somewhat exercised and when they found that testing and proving flights were going better than had been expected, they raced to be first off the blocks.

When the Comet was first over the ocean with paying passengers, the British press waved the Union Jack with vigor. It played up the Comet's speed, told how Basil Smallpiece, the airline's managing director, had balanced a penny on end for five minutes and warned that the American Boeing was twice the loaded weight of the British entry, needed a longer takeoff run, climbed more slowly, and was so noisy that it would have to take off with a light load of gas and therefore be required to make intermediate stops on its way to Europe. The Sunday *Times* of London quoted a redcap at Idlewild Airport who said, "It just shows you can't always believe what you read, sah." He referred, the *Times* man pointed out, to the "immense 1½ million dollar advertising campaign which an American airline had rashly decided to launch to tell the public that its Atlantic jet service would be first." The writer was gleeful over Pan American's difficulties with the Port of New York Authority. The Authority had ruled that the new jets could make no more noise than a Constellation when passing over the homes adjacent to the airport. "The reason for the failure of the 707 is not hard to find," the *Times* went on, looking into its remarkably clear crystal ball. "Although it weighs much more than the Comet when fully loaded, it has engines of not very much greater power. . . . If the 707 has to call at Gander it will have lost the game all around."

The first Boeings off the factory line in Seattle were medium-range versions, suited for the run from New York to San Juan or San Francisco to Honolulu. The long-range Boeings would not be ready until the summer of 1959. Although it had known for years that the jets were on the way, the Port Authority in New York, for all its noisy talk about noise, hadn't completed the jet-length runways by the time the first Boeings were ready to take off. Juan Trippe, Pan American's president, brought a 707 down to Baltimore, which had a long runway, filled his plane with publishers, and flew it to Brussels nonstop. That flight took a full hour longer than the Comet's flight to London and, worse yet, it stole the march on Pan American's own press flight which came a week later. The trip for working press took off with a full load of 110 writers from the short runway at Idlewild and got as far as the Azores before it had to put down for fuel. By the time it landed in Paris there was no record and no story. The wife of one of the correspondents, who had flown over the same night on the regular piston-engine commercial service, beat her jet-flying husband to Paris. On the way back the press plane stopped at Iceland, but even with that pause the jet, which had left Paris shortly after lunch, arrived in New York shortly before dinner.

The paying customers finally got a chance when Pan American sent its first commercial jet clipper into the air on October 26. The flight was a sellout and Mr. Trippe addressed the 111 passengers, acknowledging, with some chivalry, that the British jet had indeed been the first across the Atlantic. He recalled Lindbergh's flight, noted the progress since 1939 when Pan American sent its first clipper to Lisbon, marked this milestone, and suggested that if the jet age were to make friends of the Russian and American peoples, there would never be a World War III. The plane left at seven o'clock, but there were some disappointments. Once aloft, Captain Sam Miller announced the jet would have to make a stop at Gander. And the Paris–Rome leg of the flight, which had been planned, was canceled because the Italians were demanding a surcharge for a jet landing, thus requiring an additional fee from passengers. The first passengers paid $505 one way in deluxe class and $272 in economy class, where seats were stretched three abreast.

The race between BOAC and Pan American was academic anyway since both the Russians and the Czechs, flying the Russian Tupelov 104, were already operating jet service. Aeroflot, the Russian line, was streaking from Amsterdam and Brussels to Moscow in about three hours, from Paris to Moscow in four hours, from Moscow to Prague in two hours and a half, and from Moscow to New Delhi in six hours and a half. The Czech line also maintained service to Moscow from Prague and from Prague to Cairo, a trip it had sliced to about four hours. Both lines had been flying jets between Prague and Moscow for two years before the British and the Americans flew the Atlantic.

With the link-up that was available in the late fall of 1958, it would be possible to fly from New York clear to New Delhi by jet. No jets were as yet on the Far Eastern runs, but round-the-world trippers would shortly be able to pick up a jet again on the West Coast of the United States for the cross-country flight to New York. American Airlines, on a practice run the same day of Pan American's inaugural Atlantic service, had crossed the nation in 4 hours and 43 minutes. In New York, passengers would be able to fly to Miami on jets that National Airlines leased from Pan American for the busy south Florida winter season of 1958–1959.

Of the three jets flying, the Boeing was the largest and in first class, where the seats were two abreast, it carried a roomy lounge. The Comet was considerably smaller and rather cramped and its small doorway required passengers to stoop over when getting in and out. The Tupelov, a converted bomber which had been rushed into civilian service to steal the propaganda glory, was simply decorated and noisy in flight. The classless society had divided it into 16 first-class seats in a forward cabin and 70 tourist-class seats arranged three abreast in the rear. Between the two classes was a large galley which dispensed plain but edible food. While the plane cruised at 510 miles an hour, the passengers could amuse themselves at the magazine racks. The articles in the periodicals alternately wooed the travelers with poetic descriptions of Russia and insulted them with lacerating attacks on the West.

As far as home-consumed propaganda was concerned, the Russians were scoring well. While flying their Tupelovs into Moscow Airport and parading some new four-engine turbo-prop planes, they were limiting the western airlines that flew in Moscow to piston-engine planes. Thus for some time to come it was easy for a Russian citizen to infer that the mother-land had stolen another march by producing jets while the westerners were still flying obsolete aircraft.

The jets brought untold problems, among them traffic control and noise. Although their impending arrival had long been heralded, airports all over the world were caught with obsolete and even decaying facilities. Only two years before, airport officials were shrugging off warnings that runways would be too short. The airports in New York and London were hopelessly outdated and Los Angeles was a clapboard disgrace. Strikes threatened over the number of pilots that should be carried. Three weeks after its auspicious beginning, BOAC, hampered by labor problems, had only completed two flights in each direction. And then there was the financing. America's major carriers had ordered $2 billion worth of new equipment. How would it be paid for?

On the old piston planes that flew across the ocean, there was at least time for a night's sleep if indeed one could sleep on an airplane. The first jets all left at night and arrived in Europe in the early morning, but the passengers had only been aloft for six or seven hours. Much of that time was taken up in a lengthy dinner which began with cocktails and ended, long afterward, with brandies. A dinner like that could often take two hours or more. That left precious little sleeping time before the European dawn began to shine through the windows, and the stewardesses snapped on the lights in the cabin and started to serve breakfast. Arriving in Europe exhausted, passengers were trundled off to their hotels only to find, at 8 or 9 A.M., that the rooms they had reserved were still occupied by the previous night's tenants.

Perhaps it was the speed and perhaps the exhilaration, but

the arrival of the jets, with all their attendant problems, proved to be the beginning of a new life for the nations' airplanes. It was a new life for the traveler, too, for now he was off on adventures that took him into such obscure places as Samarkand in Soviet Central Asia, to Tashkent, home of the Uzbeks, and by propless plane into what had once been the remoteness of India.

Flying the jets into New Delhi, the new traveler could buzz off to the Taj Mahal or to Jaipur, or explore South India before flying off again to explore Singapore, go south to Jakarta and thence to Bali. Bangkok was a wild dream of gilded temples and mosaic figures of delicate people paddling canoes along the murky klongs. And then there was Hong Kong, suddenly revealed as a capsule of China, immensely colorful, brim-full of bargains, and critically short of hotels. Tokyo had shaken off the war and the occupation and was a bustling city with new restaurants, a new outlook and a dozen plans for new hotels.

The jet age unleashed a tremendous burst of energy, particularly in the Orient. It started a new vogue in travel for round-the-world trips. "Where are you going this year?" one asked one's friends in the spring. And the snappy answer would be, "Around," which was in-talk for "around-the-world." Light-hearted friends might answer, "Around the world. Next year I'm going someplace else."

BOAC announced that it would begin "jet-powered" service around the world by April of 1959 using their Comets and turboprops, or jet-powered Britannias. Qantas, the Australian Line, would have jet round-the-world service on its Boeings by July. The trip would take 72 hours. It was 434 years since Magellan made the first trip around the world. It took his expedition three years and he lost his life en route. Only eighty-six years had passed since Jules Verne's book *Around the World in Eighty Days* had appeared in Paris. It was then considered science fiction.

Now spurred by the jets and a movie version of *Around the World in Eighty Days*, a whole new class of people evolved: fancy people, adventuresome people whose ancestors had been called Café Society. The new group included all the world's playgirls, the polo-playing millionaires, starlets and expensive hairdressers, debutantes and socialites of two continents. They became known as the Jet Set and "jet" became a verb. The Broadway columnists talked of those who jetted down to Rio for three days of Carnival, or jetted over the Munich for the Fasching. The same people jetted to Paris to open the racing season at Deauville, jetted to St. Moritz to ski for a week, jetted to Rome for the Olympics. Cleveland Amory in *Who Killed Society?* blames the airplane and indicts it for murder in the second degree "with design to effect death but without deliberation, meditation or malice aforethought."

It was noticed—but in the flurry of the jets perhaps only in passing—that the French had written off the proud old *Ile de France,* a symbol of the good rich life of the late 1920s and the 1930s when, if you went to Europe, you went by ship. The French Line consigned her, savior of the sea lanes (she had taken 753 survivors from the sinking *Andrea Doria*), to a Japanese scrap merchant. In a last-minute reprieve she was rented by a motion-picture producer for the film *The Last Voyage.* At picture's end she was blown up and left to burn for the glorious action-packed finale.

Besides the loss of a famed old ship, there would soon be the loss of a famed old playground. Castro had marched out of the mountains and taken control of Cuba. He was avowedly no Communist. He was eager, he said, to make Havana even more popular for visiting Americans than it had been before. He had big hotels to fill, several of them completed as the revolution was under way, among them the big flashy Riviera and the tall Habana Hilton. What was more, the huge World Travel Congress, staged annually by the American Society of Travel Agents, was scheduled to hold its autumn conclave in Havana. The presence of so many travel agents and so many travel executives from America and from all over the world had been known to bestow benefits on whatever country in which the agents had chosen to meet.

All that summer of 1959 American newspapers and magazines carried huge ads that shouted, "Go carefree go Cuba." They insisted, "Cuba is gayer than ever," that "Cuba's new democracy in action offers you the new freedom, new fun and the biggest welcome in the whole Caribbean." One by one the directors of ASTA, the travel agents' group, were brought to Havana and romanced. So was the influential editor of *Holiday*, the American travel magazine. *Holiday* ran a story about the new Cuba with full-page pictures of such local luminaries as Señor Bacardi and Señor Castro himself. It didn't matter that the ASTA convention had been inherited from the much-hated Batista regime, whose members were being paraded to the wall. The hotels of Havana and the beaches of Veradero had been empty since New Year's Day, when Castro took over.

ASTA decided to hold its convention anyway and it started well. The first arrivals found the airport decorated with bunting. Bearded soldiers with submachines lurked in the corners, but the delegates were pushed through customs so fast they hardly had time to look. Gifts of rum and cigars came flooding into the delegates' rooms. Castro himself appeared at the first gala, took off his pistol belt and waded into the sea of travel agents. Seemingly awed by his bearded presence, they crowded him as if he were a movie star, took their delegate's badges off and elbowed each other to have him sign his name across the face. Castro Wows Delegates was the headline in the convention's daily newspaper.

After the reception Castro was due at a press conference at half past ten. A hundred and fifty pressmen waited for hours and finally his lieutenants began to vamp for him. They told of their plans: Santiago, center of the revolution, would be turned into a great tourist city with a jet airport with direct service from New York by 1961. Nearly half a million dollars had been scheduled in advertising in American periodicals before the end of the year, a million and a half the following year. The landing tax would be removed. Every tourist would be given twenty-five stamped postcards on arrival and a Cuban hat filled with fruit on departure. While his aides were telling this dream-filled story Castro, it turned out later, was in a television studio flaying the United States.

On Wednesday of that week, with most delegates visiting resorts and historical sights (the Spanish-American War was now being called "the intervention"), a counterrevolution plane suddenly appeared, showering Havana with leaflets. Simultaneously, a car raced through the streets dropping explosives. Two people were killed, forty-five were injured and a Cuban plane that rose to give chase was hit by friendly fire from the ground. Castro was in Camagüey arresting an old friend, Major Hubert Matos, but he returned that evening and attended a convention reception at the Habana Riviera Hotel. That night shouting mobs were in the street in front of the showpiece hotel, where many of the delegates were staying. The next morning the revolution's newspaper fanned the fires with a screaming headline announcing that the "planes" had come from the United States. In Castro's second tirade over television the following night, the plane that had become planes, now turned into bombers. Such an outrage, he screamed, could only be compared with Pearl Harbor and the blowing up of the *Maine*.

Despite all this, when Castro showed up at the final convention meeting he got a plaque from the ASTA executives for his help in preparing such a successful convention. In turn he invited the officials to a private lunch, then didn't show up. He was on hand for the final banquet and, incredibly, the ASTA president awarded him the only honorary membership in the society's history. A week after he had been mobbed at the first reception, the only honorary member of the American Society of Travel Agents got out of a helicopter carrying a Belgian submachine and, addressing 300,000 followers, he ranted against foreign vested interests, demanded independence for Puerto Rico, deplored the "bombing of Cuba" and shouted for the return to the firing squads that had already dropped 450 Cubans. Two years later the ASTA delegates, gathering this time in the cultural center of Las Vegas, huddled around television sets while President Kennedy told of the Russian missiles that had been discovered in Cuba, placed there with the connivance of the travel association's only honorary member.

The trouble in Cuba sent ripples all over the West Indies, affecting every island from Jamaica across to Puerto Rico. Gradually a calm descended, and as the exodus from Cuba continued, both Miami and San Juan reaped the rewards. The famous restaurant and nightclub operators of Havana had fled, too, bringing their talent and their style under the protective stripings of the American flag.

A year after the Jet Age began 1.5 million tourists went abroad spending $2.3 billion. For the first time in history more of them flew the Atlantic than sailed it. This age at last belonged to the jets. The average overseas traveler was forty-six, and if he went by air he stayed away 41 days, spent $1440; if he went by sea he stayed 68 days, spent $1710. There were now 3000 travel agencies in the United States ready to take care of all the details, to sell a Sam Snead tour for golfers or a Charles Goren tour for bridge players. There were gourmet tours, garden tours, theater tours and single people's tours. In fact, there were so many Americans abroad that sociologists looked at Europe's supermarkets, its self-service restaurants, the penchant of its youth for blue jeans, the universal appeal of Hollywood's movies, and wondered if the world wasn't suffering from Americanization practically everywhere.

In 1960 Americans went to Europe to attend such diverse assemblies as the International Atom Seminar in Berlin, the International Congress of Beauty Care and Cosmetology in Amsterdam, the Planned Parenthood Federation's meeting at The Hague, the International Congress of Dental Surgeons, the International Geological Congress, the World Congress of Widows and Widowers and, at Purerend in the Netherlands, an International Reunion of Children Born in Leap Year.

Trips to attend some of these meetings, if they could be traced to a reasonable business need, were, in part anyway, deductible on one's income tax. But that was only a fragment of the reason that inspired so many Americans to travel. All sorts of experts looked for the basic motivation. Said Nancy Mitford, "The North Americans very naturally want to get away from North America. They are also after their own origins. Although they descend from people who could not succeed in Europe and furiously shook its dust from their feet, they have a sentimental feeling for ancestors."

Closer to the true reason was the simple fact that more Americans than ever before had money and time. There was new wealth and new leisure. They were curious and travel to

Europe was easy and, for budget-watchers, not very expensive. But the great rush to Europe, which had been so heartily encouraged by Washington in the years after the war, took an odd and serious turn. It was discovered that Americans were spending $1 billion more in foreign travel than foreigners spent here. In one September week of 1960, $181 million in gold eddied out of governmental coffers. *Time* magazine called it "the biggest weekly commercial loss since Depression 1931." The nation was suffering a serious gold problem. Ironically enough, the biggest drop in gold came during the height of what President Eisenhower had proclaimed as Visit the U.S.A. Year, a year designed to bring foreign currency to the United States. Unfortunately, a committee appointed by the Secretary of Commerce to promote the year immediately set up three subcommittees, adjourned for the day and never met again. The director of the International Travel Office of the Department of Commerce complained he couldn't even get a U.S.A. poster during Visit U.S.A. Year. What promotional material was mustered was sent to U.S. embassies and consulates abroad. When one such package arrived in Brussels, the U.S. embassy called in Arthur Haulot, head of the Belgian National Travel Office and of the European Travel Commission, which had done so much to spark the return to commercial tourism on the Continent. Asked for advice, a perplexed Haulot reminded the embassy that it was the United States itself which had urged the foundation of the European Travel Commission, and which had sent its advisers to rehabilitate the European travel industry at war's end.

The first Visit U.S.A. Year was a flat fizzle. The increase in foreign spending for the first half of 1960 indicated a gain that was less than the year before. Indeed it was the lowest gain in all but one of the previous six years. It failed because Americans, so expert at selling soap, underarm deodorants, and breakfast cereals, proved totally inept at selling themselves. It fell flat because United States visa requirements were too difficult and too insulting. Even the Russians didn't ask the kinds of questions demanded on a visa for the United States. It fell flat because the members of America's tourist plant were insular and provincial. The Tourist Commissioner of Virginia averred that he was gearing his promotion to "English-language tours for English people." That way, he said, he would get into no language problems and no segregation problems. All that talk about international understanding was simply too high-flown, said one tourist director of a western state. He was wondering how he could ship his travel folders abroad. It fell flat because the United States needed travel offices abroad, of the type foreign countries long ago established in the United States. A bill to create a United States Travel Service died in the 86th Congress because legislators, in the words of Congressman Robert Hemphill, thought setting up travel offices abroad meant "more government," and he was against too much government.

Rather than promote visitations from foreigners to offset our dollar drain, some pundits—Arthur Krock of the New York *Times* was one—called for dropping the duty-free limit on foreign tourist purchases from $500 to $100, and suggested that the government curtail funds which American citizens could take out of the country. His view did not altogether agree with the opinion offered by the *Times*'s own editorial supporting a strong article in the *Saturday Review* which had urged a three-pronged crash program: a vigorous drive to attract foreign tourists, an improvement of facilities for foreign visitors and a relaxation of entrance formalities.

Eventually the duty-free limit for returning American tourists was dropped from $500 to $100, not a very significant gesture since the wealthy travelers who wanted to buy abroad would simply pay the duty. The incoming Kennedy administration had paid little attention to a survey made in 1959 which showed that only 10 percent of midsummer travelers were coming home with more than the $500 limit. Eighty-two percent spent less than $200. A few years later the belt was tightened further when the customs limit was dropped to $100 of the retail value of the purchased articles. Unaccompanied articles didn't count. Reprehensibly, the bourbon lobby in the United States was able to tack a rider on that bill which dropped the tax-free limit from five bottles of liquor purchased abroad to one bottle. Here again the American traveler was duped and he hardly uttered a squeak of objection. It clearly made no difference in the dollar drain whether the bottle of scotch purchased by an American was bought tax-free abroad or retail at his corner liquor store; the amount of dollar loss to the government was exactly the same. And certainly it could not be concluded by anyone with a straight face and a clear mind that the imposition of such restrictions on liquor bought abroad would cause the buyer to change his taste and commence drinking homegrown bourbon because his scotch or tequila was no longer duty-free.

Despite the shortsightedness of the people's representatives on Capitol Hill, however, a bill establishing a United States Travel Service was finally pushed through Congress. Secretary of Commerce Luther Hodges appointed Voit Gilmore, a North Carolina politician, motel owner, ex-naval officer and onetime Pan American employee, as its first director. Gilmore organized a headquarters in Washington, set about opening offices in Europe, South America and the Far East. Advertisements to come to America began to break in foreign journals, and the program to bring tourists to the United States was at last underway. At home, meanwhile, urgent campaigns were launched to simplify our visa requirements and our customs and immigration handling at our docks, airports and borders. Domestic transportation companies were urged to develop special rates for foreign tourists. Thus was born the 99 days for $99 bus ticket, and an all-inclusive ticket on feeder airlines that offered 21 days of flying for $150.

Still Mr. Gilmore ran into heavy sledding, first from the foreign press, which pecked at him for what was called "the dollar curtain," meaning our prices were too high; and secondly, for our visa barriers (which were being modified); and finally, for our general ineptitude in handling foreign visitors. His problems multiplied when he trudged up the Hill asking for a budget of $4.2 million for fiscal 1963. He was set upon by committee members, notably Congressman John J. Rooney of Brooklyn, New York, a celebrated xenophobe who submitted Gilmore to an unrelenting inquisition on the money he had already spent as director. At some government expense Rooney had sent a pair of sleuths to the United States Travel Service headquarters in Washington and then across the world to investigate the Department's procedures. Now, he wanted to know the answers to such burning questions as: Why was foreign champagne served at the opening of the Tokyo office? Why was the rug for the Tokyo office bought in Hong Kong when we have mills that make rugs in New York State? The answers were that French champagne had been donated for the opening and that it was cheaper to buy a rug in Hong Kong than it was to have one shipped to Tokyo from New York. The answers proved immaterial. The budget of the United States Travel Service was cut to $2.6 million. For a nation with an acute dollar loss competing with Britain's $5 million budget, Canada's $6.5 million budget and the $4 million being spent by Mexico and France, the American budget was lamentable, embarrassing and bad business. Even such nations as Ireland, Greece and Italy were spending more on tourist promotion than the United States and they were making it pay. Frustrated by his excursion in Washington, Gilmore resigned and his assistant, John Black, took over. Black's chances hardly looked better. Although earnings from foreign tourists had jumped a healthy 20 percent, the $480 million spent by visitors in the United States were a mere third of what Americans were spending abroad. Nor was the imbalance offset by a reference in President Johnson's balance-of-payments message in which he said that the time was not proper to encourage travel abroad, but to encourage travel within the United States. There were continuing rumors, always met by strong dissident editorials and subsequently by government denials, that Washington would act to curtail foreign travel by its citizens. There was considerable speculation that a tax on foreign travel would be imposed. Its opponents formed a heavy defense, invoking the Constitution, deploring the limitations on free concourse of peoples, warning that a tax would prove discriminatory since the rich would merely pay it and only the poor would suffer.

Jackie Kennedy had roamed the world and made it smart to visit Greece and eat in chic, elegant restaurants in Vienna. As First Lady, Lady Bird Johnson organized canoe trips through the West, pushed beautification campaigns in the United States and so subtly—some thought not so subtly—implemented her husband's See America drive. There was some help from the great fairs—Seattle in 1963 and New York in 1964–1965. But Seattle was relatively small and lacked the competitive excitement of Brussels in 1958. New York was an over-commercialized bust that failed to engender national, much less international, zeal, failed to draw the crowds and failed to pay back its investors. It would be Montreal's turn in 1967.

Despite the presidential supplications, the competition to go far and see the exotic was almost overwhelming. American Express was taking full-page ads in newspapers which announced, "We hate cash." It urged Americans to buy traveler's checks or just take a trip and charge it on an American Express credit card. Diners Club spread a vast network around the world, affording its members the opportunity of charging a meal eaten in San Filieu de Guixols or a string of pearls purchased in Tokyo. Africa was unveiled and Americans who had never been farther than their hometown zoo on Sunday were flying into Nairobi, outfitting themselves with bush jackets and safari boots and riding off to spend a night at Treetops to watch the animals from an upholstered perch on a limb. They slept under canvas at Amboseli, where the hyenas laugh eerily at night and the giant rhinoceroses roam the dusty tracks by day sharing the real estate with the elephants, the herds of gazelles and zebras and the giraffes.

Ethiopian Airlines, helped by TWA, pushed a vigorous program to bring tourists to the Land of the Queen of Sheba, as they called it. South Africa, its racial edicts too inhuman for the tastes of most Americans, fought a losing campaign for travelers from the United States. Zanzibar was a colorful port of call, but its politics went too far the other way, and after ousting the sultan it sought alliances for awhile with both Communist Cuba and Communist China. West Africa suffered the disadvantages of heat, bitter memories of the Congo and unstable governments that in the mid-1960s toppled one after another.

But there were other lures. Japan, almost alone among nations, had made marvelous improvements in railroading. A streamlined twelve-car electric-powered train was streaking along the 312-mile route from Tokyo to Osaka in just 3 hours and 12 minutes. In New York the helicopters were flying to Idlewild, newly renamed Kennedy Airport, from the top of the Pan American Building in midtown. In Hawaii helicopters offered perhaps the world's most exciting sight-seeing ride, skimming over the knife-edge ridges along the Na Pali coast of the island of Kauai, dropping into hidden valleys and depositing vacationists on remote beaches to which no roads led.

In 1963 Hilton opened hotels in Holland, on a hilltop just outside Rome, in Tokyo and in Hong Kong. The huge inn built in the British Crown Colony was its most ambitious, for it was an expansive structure with several decks of shops, a rooftop restaurant, a cellar nightclub, an outdoor swimming pool on a setback, and a magnificent cuisine unmatched in the Orient. It rose high above the harbor on the Hong Kong Island, where, with the equally elegant Mandarin Hotel, operated by Pan American's Inter-Continental Hotels, it created a new vogue across the harbor from Kowloon, which had been in favor ever since Hong Kong's sudden postwar rise to tourist favor.

Later Hilton opened a new hotel of considerable style and elegance in the Kahala section of Honolulu. It was called, to be sure, the Kahala Hilton and until a porpoise took an annoyed snap at a tourist, the guests were invited to swim with the dolphins in a hotel-side lagoon. Sheraton had invaded the gorgeous four-mile stretch of beach at Kaanapali on the Hawaiian Island of Maui, to be joined by two other hotels and hints of more to come. Waikiki became a canyon of hotels and office buildings and some that combined both offices and hotel rooms. The airlines dropped their air fare on "economy" class—called "thrift" class, now—to $100 between California and Honolulu, and Hawaii's tourism knew no seasons.

Inter-Continental Hotels opened a rakish hostelry on Sadie Thompson's island of Samoa and survey teams moved on into Tahiti, which was now connected to Los Angeles by nonstop jets, and into Fiji, where the last cannibal had eaten the last European in 1896. A new group of rising hoteliers, last heard from in Florida, planted an elegant hotel called the Regency on New York's Park Avenue, and it immediately attracted such clients as Princess Grace and Prince Rainier; Elizabeth Taylor, then newly betrothed to Richard Burton; and most of the

monied classes of Hollywood and South America. To provide
something for everybody in the big city, the group, known
as Loews Hotels, also put up the gargantuan Americana on
New York's West Side, as well as a midtown motor hotel.
Hilton topped the Americana with the New York Hilton,
which could offer a duplex rooftop suite complete with movie-
land fittings and a view of the surrounding real estate so stunning at night that Manhattan appeared as a bejeweled fairy-
land.

Hilton joined Sheraton with a hotel in Tel Aviv, a neat feat
for the Hilton chain, which was also operating in Cairo. Then
in the spring of 1966 and in his seventy-eighth year, Conrad
Hilton came to Paris, opening the first major hotel in the
French capital since the George V had been unveiled seven
years before the Germans marched into the city. While it
nestled in a residential neighborhood at the feet of the Eiffel
Tower and was not at all near the habitual tourist centers, it
offered such unusual gambits, at least for Paris, as a coffee
shop dispensing waffles and pancakes (on the opening weekend the waffles were being served with a slice of raw onion
and a dill pickle, but that got rectified), and a restaurant called
Le Western. Amid a decor of white saddles and gold-sprayed
stirrups and the music of a rinky-dink piano, Le Western
dished up pinto beans, corn on the cob and cowboy steak
speared on a ranch knife. The French were fascinated.

On the high seas the Swedes brought in their fourth new
Kungsholm since 1923. It had gyms, saunas, indoor and outdoor pools and beds 6 feet, 7 inches long and 3 feet wide.
Every cabin had its own bath. In the air, comfort was being
sacrificed for speed. The flying berths were long since gone,
even on the eleven-hour flights from California over the Pole
to Europe, from New York to Athens and Israel, and from New
York to Buenos Aires. On some lines even the forward lounges
in first class had been removed, allowing nothing more than a
short walk down the aisles for a change of scene or a bit of
exercise. The latest word was "stretch"—fashioning elongated
versions of standard planes. First to be stretched was the
DC-8, which in its new versions was called the DC-861, the
DC-862 and the DC-863. The 1961 model measured 36 feet

and 6 inches longer than the old DC-8 jets. In the version bought by United Airlines for its California–Honolulu run, three movie screens were needed for all to see the film.

Planes were getting longer and soon they would be getting even faster. Ships were made even more comfortable. Cruises were longer and more luxurious. From the Aeolian islands off Sicily to Macao in the South China Sea, slow-moving small craft were being replaced by hydrofoils that rose on runners and sped over the tops of the waves. Hovercraft were being introduced that could rise above the water and then rocket across it at new speeds. Ads beseeched Americans to "Take time, take the ship." Others suggested that "Air France is for gourmets who can't wait to get there."

"It's Life Aboard the S.S. *France* that makes the difference," crowed the French Line, while on another page another message cast doubt. "Why spend it all getting there?" the ad said. "Fly Icelandic, lowest fares to Europe." Japan Air Lines stressed its hospitable ways, Alitalia pushed its tours (Portugal and Madeira, 15 days, $465.50 and Portugal, Spain and Morocco, 21 days, $599). The far places whispered seductively. Under the lovely photograph of an Indian beauty the catch-line read, "No wonder there is a Taj Mahal." And from the Balkans, with the help of an imaginative copy writer on Madison Avenue, Greece beseeched anyone with the price of a ticket to "Hear the goatbells. Smell the jasmine. Taste the yoghurt and honey."

Travelers would go wherever the beautiful people turned up last year. The Jet Set became, as *Venture*, a svelte (though short-lived) travel magazine, called them, the Placesetters, and the Placesetters settled into Sardinia and bought property along the Costa Smeralda, where the Aga Khan was developing a resort. They were back in Deauville, where a young French count who had been elected mayor was drumming up a renaissance of the gay times in Deauville before he himself was born—the days when the Dolly sisters were the Gabors of their time and kings bought palaces for soubrettes. In the mid-1960s the Placesetters suddenly discovered Acapulco, which Hollywood had discovered thirty years before and the not-so-beautiful people had enjoyed for the last decade. Suddenly it was revealed that Gloria Guinness had a house at Las Brisas, a lovely cottage colony with private swimming pools hanging on a cliffside just outside Acapulco. It had been perched there for nearly ten years and had always enjoyed a faithful following. But when Mrs. Guinness left Palm Beach in her private jet and headed—not for her private homes which she had in New York, Paris and Deauville—but to Acapulco, the sunswept strand exploded all over again. Guests at Las Brisas in the recent renaissance included Frank Sinatra and Mia Farrow, and those other amours, Lynda Bird Johnson and George Hamilton. In its housekeeping villas lower down on the slope where Gloria Guinness had *her* home, the tenants included Texan Clint Murchison, Baron Guy de Rothschild, presidential aide Robert Kintner, publisher Samuel Newhouse and erstwhile premier Anthony Eden. Of all of Acapulco's hotels the fashionable stayed at only two, Las Brisas and Villa Vera, a tennis club operated by Teddy Stauffer, a onetime bandleader whose past fame included a marriage to Hedy Lamarr. As a turncoat from Palm Beach told a New York *Times* reporter one day, "I'd known about Acapulco's beauty and incredible weather for years, but then I found that if one knew the right places and the people here there was no need to stay in hotels or mix with Cedar Rapids corn-syrup manufacturers." As soon as the beautiful people got bored, they would find a new place and fly off—by jet.

Above all it was the jet that had created the age. As the first decade of planes without propellers drew slowly to a close, it became clear that it had been a time for explorations, a time for adventure. It was a time when adventurers of moderate means walked the corridors of far-flung museums, penetrated jungle trails, bargained in the back alleys of Kowloon, rolled down Sussex country roads, gaped at the walled city of Dubrovnik, rowed a boat on the Sea of Galilee, sent home postcards from Kyoto—experiences which twenty years before would have been too fanciful even to have been the stuff of dreams.

In the future there could only be faster planes, visits to China, a vacation in Antarctica and a trip to the moon. What else was left?

II (1970–1980)

THE TRIP TO THE MOON was accomplished—though by astronauts, not tourists. The first timid advances were made toward China when an American ping-pong team accepted an invitation to travel to the land that had been closed off to U.S. citizens for a score of years. Then came Henry Kissinger's secret trip followed by the highly hyped excursion by President Richard Nixon, who arrived in Peking accompanied by his wife, his aides (many of whom were to be in jail before the decade was over) and a glittering press corps. The whole show appeared in living color in every home in America that had a television set, and it stirred a new wanderlust for the Forbidden City and the vast land that surrounded it. Ultimately there would be trips arranged by the China Travel Service. One of the first agencies to be involved in sending Americans to China was Lindblad, which was already dispatching visitors to Antarctica, where they equipped them with orange-red jackets and buzzed them off to ice floes to commune with penguins. That took care of the moon, China and the southern ice.

The 1970s, at the outset, gave all sorts of promise, especially in new and fanciful ways to travel that would make journeys incredibly shorter and vastly more comfortable. The anticipated appearance of a plane that would fly faster than sound, that is, at a speed over 700 miles an hour, created a national debate in the United States as well as a rumble in Britain and other countries. The new mode of air travel, dubbed the SST (for supersonic transport), was already being flown by major military air forces. Britain and France had entered into an entente to build a joint commercial SST to be called Concorde. The Soviets announced their version—a TU-144, which would conquer the immensity of that vast land. It was tested before the 1960s were over.

Both the British-French version and that of the Soviets were to be planes of limited speed, size and payload. Concorde would cut the time of standard jets in two, flying at about 1400 miles an hour, or Mach 2 (twice the speed of sound), and would carry 128 passengers. TU-144, slightly faster, would carry fewer people—a total of 120.

The United States, however, was planning a different sort of airplane, one with a titanium body that would permit it to fly higher and faster and carry nearly three times as many passengers as the other two. It would cross oceans at 1786 miles an hour with nearly 300 passengers. Boeing would build two prototypes financed by the United States Government.

But the proposal appeared at the very apogee of the environmental movement, and the expenditure of such vast sums, for what appeared to many as a questionable purpose, provoked hysterical partisan battles that had not been seen since the introduction of the horseless carriage a little more than seven decades before. It was alleged that the sonic boom would prove unbearable, would deafen passengers on ships, upset the life-cycle of all the fish in the sea. Flying this airplane at supersonic speeds across land was unthinkable. Hadn't those military SSTs broken windows all over the West when they practiced high-level dogfights?

It would waste fuel, pollute the air. Moreover, it was darkly rumored—and that unassailable patriarch of the airwaves, Walter Cronkite, had even repeated it on the air—that supersonics flying all over the place at unheard-of altitudes would peel back the ozone layer that was protecting us all from radiation. We would suffer skin cancer. Polar ice caps would melt, causing oceans to flood.

And if that wasn't enough to give anybody the heebie-jeebies, to say nothing of an incipient carcinoma, there was the matter of noise—both sideline (when on the runway) and when taking off and landing. General Electric, proposed builder of the engines, claimed the takeoff and landing noise would be well within the limits of existing jets. It was acknowledged that sideline noise would have to be reduced.

Then, the opponents said, if the pollutants and the noise didn't get you, the economics surely would. Why should all that money be poured into a toy to be manufactured so the rich could fly faster? Why wasn't the money being put, instead, into education, medicine and environmental control? Or mass transit itself? The New York *Times* cleared its throat yet once more and wrote,

> Against the SST in the beginning and against it now is the threat of unbearable noise, both from the sonic boom and the deafening sound at take-off. The sum earmarked in the transportation budget for this needless and noisy monster is almost $300 million. Mass transit is down for $80 million which it may not even get. An administration that cannot order its priorities better than that should have no complaint if Congress offers to help.

On the other side of the battle line were those who worried about the loss in balance of payments should the new technology be abandoned by the United States, the world leader in the air frame construction, and be picked up by the European Concorde, or even the Soviet SST. Senator Henry Jackson, representing the State of Washington, home of Boeing aircraft, opined that the SST appeared to the public as a manifestation of the military-industrial complex, a combine that in the summer of 1970 provided a fat target for abuse. Even President Eisenhower had warned about such a partnership. There were those historians who recalled that in 1865 the British had passed the Red Flag Act, which demanded that all steam carriages be required to carry a crew of three, one walking at least sixty yards ahead of the vehicle and waving a red flag. The speed limit was set at two miles an hour inside city limits and four miles an hour out of town. The law remained on the books for thirty years.

Many who saw SST in a favorable light asked, were we not succumbing to the same sort of overreaction that traditionally met each new form of transport? Was it not Napoleon who said to Robert Fulton, "What sir, you would make a ship sail against the wind and currents by lighting a bonfire under her decks? I pray you excuse me. I have no time for such nonsense."

But the mood of the nation, aroused by fear, was to prevent the Senate (though not the House) from passing the prototype bill, and America abandoned the age of supersonic transport to the British, the French and the Russians. At the time, the United States had 26 million citizens living below the poverty level, and now those in the air frame industry in the Northwest and in California, and those in scores of plants manufacturing parts for the proposed aircraft, were plunged into deep gloom. There would be, so it was foretold, wholesale layoffs as well as a further loss in the balance of payments.

The anti-SST people, having won a whopping victory in America, turned their guns on the British-French airplane, which in short order would attempt to begin service on the lucrative North Atlantic run between Europe and the United States. Concorde was put through a vigorous program of tests.

In the early autumn of 1973, with thirty-one other passengers, I flew the first eastbound transatlantic test flight from Dulles Airport in Washington to Paris. The Concorde people had been required to use Dulles because it was under federal control and the administration was in favor of supersonic transport, while New York had vowed that this aeronautical monster, with all its built-in hazards to mankind, was not about to drop down in lily-white Gotham.

It was a somber moment. The French ambassador was on hand to wish us well. On board were eight tons of computers manned by a special crew. (When the Concorde was finally certified, that equipment would be replaced by passenger seats.)

At 7:46 in the misting Virginia morning we moved to the runway. Inside the cabin the rush of air from the engines was similar to the noise of driving one's car with the window open on a blustery day.

In forty-one seconds we were off the ground. Inside the cabin the noise didn't begin to compare with that of the DC-9 that had brought me from New York to Washington the night before.

We turned east and in fifteen minutes we had left America with a heading south of Atlantic City, south of Boston and

Newfoundland, following the circle that would bring us over Mont-St.-Michel between Normandy and Brittany, then Chartres, the cathedral city outside Paris.

At 8:10 we left the American coast, got permission to open the throttles to Mach 1, the speed of sound. A computer indicator flashed the numbers on the wall of the bulkhead. At 8:12 we were flying faster than sound and in twenty-six minutes we had hit Mach 2. Now we were speeding more than 1400 miles an hour ten miles above the Atlantic. There was no sensation beyond exhilaration; the passengers—including Harding Lawrence, head of Braniff International, and his wife Mary Wells, and Pierre Cot, then president of Air France —were nibbling on omelets and *pâté de grive* served with champagne.

By the time we were two hours out of Washington we had begun to slip between Land's End in England on one side and Brest in France on the other. The wheels nestled to the runway at Orly Airport in Paris exactly 3 hours and 33 minutes after lift-off in Dulles. There came a great burst of applause in the cabin. As I wrote at the time, nobody ever made more points with one touchdown.

That night Concorde officials told me of the campaigns in Britain (though not in France) to knock SST out of the skies. An elaborate staging had been set up. Concorde would make a highly publicized test flight all around the coast of England. The day of the announced flight came and newspapers, police and airlines got a blizzard of complaints. Windows had exploded. Pregnant cats had aborted. Horses had reared. But in truth Concorde hadn't flown that day. It was deliberately held back to assess public reaction. Sometime later, unannounced, it made the series of flights first proposed. The public reaction was nil.

But despite that unmasking of the hysteria that was greeting SST, as similar opposition in prior generations had greeted the steam locomotive and the automobile, Concorde's troubles were just beginning. Although the U.S. Secretary of Commerce, William T. Coleman, authorized both Air France and British Airways to commence service through New York and Washington for a sixteen-month trial period beginning in

February 1976, the Port Authority of New York and New Jersey banned Concorde operations for six months.

By spring of 1976, Bicentennial Year in the United States, regular Concorde flights began on a three-times-a-week basis. But the Port Authority of New York and New Jersey continued a long, nettlesome and boring series of court delays that lasted into the summer of 1977. By the summer of 1979, when I next flew Concorde—this time from New York—there was no more talk of noise, of mass demonstrations by those who lived around the airport. There was no talk of sonic booms. The so-called greenhouse effect that would irradiate everybody was an idea in eclipse. Indeed, it had been determined that, if anything, the ozone layer had increased, not decreased, during the period of tense argument. Apparently the ozone layer was not affected by commercial or military supersonics. (The United States banned chlorofluorocarbons emitted by aerosol sprays, and an estimate in 1979 seemed to indicate that the ozone layer was truly being eroded by the use of aerosol sprays in other countries and by the use of chemical gases in industry and in refrigerants.)

The Concorde fare was higher than standard first class, but there were businessmen to whom the time saved was significant. Most important, SST had minimized jet lag, whose uncomfortable effect on passengers was so pronounced, especially on those taking overnight flights from America to Europe. The Concorde cuisine on Air France and on British Airways was superb, proving that it was possible to attain culinary expertise even in limited galley space and short flying time.

The age of supersonic transport had indeed arrived, but what dampened its future was not the bugaboos that so many alarmed citizens had seen in it. Rather, the small size of the Concorde—with its limited payload, plus the hugely increased cost of fuel caused by OPEC and such political developments as the end of the Shah's reign in Iran—continued to threaten the Concorde.

Despite everything, Concorde gained its adherents, and after two years it was making money on the North Atlantic run. With its Concorde, British Airways forced Pan American to eliminate its morning flight from New York to London.

Having been thus obliged, British Airways then added two Concordes a week on its London–New York schedule. Yet to come was twice-a-day service every day of the week. It was also possible to fly Concorde from the Dallas-Fort Worth Airport in Texas with Braniff crews aboard as far as Washington and British crews thereafter, the arrangement being reversed on the westbound leg.

Air France was making money on its North Atlantic crossings, too, but like British Airways it was losing on other routes. Production of the airplane, which started in late 1961, was ended by both governments late in 1979 after sixteen planes had been built at a cost of some $9 billion.

While big ships were disappearing from the sealanes, big planes were roaring down runways headed for the open skies. Boeing, which had reacted so bitterly when its supersonic transport was voted out of existence, was already deeply involved with the 747, the first of the so-called wide-bodied airplanes—so wide, in fact, that its sides were almost straight, eliminating that tube effect that had been so pronounced in the Comet, so apparent in the 707 and the DC-8 and, some said, so evident in the Concorde.

Boeing's 747 was almost 80 feet longer than the 707 that had revolutionized travel when it appeared in 1958. It could carry nearly 400 passengers, plus 40,000 pounds of freight. It had two decks. The flight deck was on the upper floor, as well as an extra lavatory and a lounge. Pan American chose to use its upper deck for first-class table service, with white napery and course-by-course service. There was an air of ship travel about this service, with passengers being paired at tables sometimes at random, but often according to age and field of interests. Other airlines elected to use the 747's upper deck for first-class passengers, devoting the vast lower deck to those buying less expensive tickets. Seats seemed to go on and on and were often divided into salons, with different movies (showing on six-foot-wide screens) playing in different cabins.

Initially, existing ground equipment wasn't geared to handle the widebodies. New stairways were required, new loading systems, new waiting rooms able to contain at least 350 passengers and new baggage systems had to be devised. Often this meant whole new airports.

The 747, moreover, was faster than the 707 and would soon render the first pure jets all but obsolete. Boeing had on its boards a new set of short-distance aircraft—727s and 737s. A 737, for example, could take 120 passengers from New York to the Bahamas in little over two hours.

Other wide-bodied airplanes came from McDonnell-Douglas, with its DC-10, and Lockheed, with its L-1011.

Inevitably the question arose about what would happen in case of an air accident involving large numbers of people. The actuaries, one airline executive figured, would indicate the loss of five jumbo jets in the first two years. That never happened, but other things did. First there was an outbreak of terrorism by political groups eager to attract attention to their causes. Jumbo jets were hijacked, forcibly flown to remote locations and blown up. One Air France jumbo that terrorists had forced to land in Uganda, where the then President, Idi Amin, gave the hijackers protection, was rescued by commandos from Israel, a daring caper that became both book and movie.

Even without terrorist help there were jumbo accidents: a dreadful crash between Pan Am and KLM on the runway at the Canary Islands airport, and several fatal DC-10 crashes. One of these DC-10 accidents, in Chicago, killed 273 persons in May 1979, grounding the airplane for a protracted period. But mishaps failed to stem the tide of the 1970s. Americans were irrepressibly on the move.

Young travelers, and some not so young, found that Icelandic Airlines, which was not a member of the airlines association and thus not subject to its fixed tariffs, was a cheaper way to travel across the ocean. Once on the other side there was Eurailpass, a coalition of thirteen western European countries when it started in 1959, offering a low-cost package rail fare. By 1980 the participating countries had risen to sixteen and the price for a two-month second class Youthpass was up to $290.

Charter flights, with one entrepreneur leasing a whole plane and flying it off to some sunny resort in Spain or Africa, had been a popular mode of vacationing in Europe for years.

The United States, heavily bound in federal red tape and regulations, started timidly. The Civil Aeronautics Board loosened the straps slightly by permitting charter flights by "affinity groups." That meant, if you belonged to an association you could fly as a group and pay less than the standard fare. It proved a ludicrous arrangement, with instant memberships in nonexistent societies being offered by freelance "travel agents" operating out of telephone booths at Kennedy Airport. Soon, regular charter flights without the necessity of an affinity group were available and the assault on airline regulations was on. It was led by Frederick A. Laker, a Briton possessed of flair, cheek and boundless energy. When a judge in a high English court called him a "merchant adventurer," Freddie was pleased. "Sir Walter Raleigh was one," he said delightedly, "a merchant adventurer in a sailing ship buying furs and spices and bringing them back to England." At first Laker added a few fillips to charter travel. He started "time charters," later known as "advanced booking charters." Under that scheme a traveler who would book and pay for a trip in advance could get a flight at a considerable saving.

Laker called such words as "charter" and "schedule" "figments of imagination dreamed up by IATA," the acronym for the association of international airlines which operate on fixed schedules. From advanced-booking trips, which landed at Gatwick Airport in London rather than at Heathrow, and which proved a great success, Laker ultimately broke the official bonds that were restraining his pet project, Skytrain. Using 345-seat DC-10s, Laker began a regularly scheduled service to London, first from New York and later from Los Angeles. Show up, buy your ticket, and if the plane is full then go the next day. Such frills as food, drink and movies were strictly cash and carry. In the Laker world there were no no-shows, no reduced fares, no waste and no thrown-out food.

Skytrain netted Laker a fair amount of booty for a merchant prince, as well as a title. He was now Sir Freddie. The renegade had been accepted by the establishment. More than that, some credit Sir Freddie with starting the flood that broke the gates of strict federal regulations that for forty years had kept the airlines laced in a tight corset of bureaucracy.

Late in 1978 Congress passed the Airline Deregulation Act, which allowed the airlines to pick up new routes, to fix new flight schedules and to set new fares. The turnaround was immediate. Airlines like Braniff, once known for its routes to South America, were flying off to such unlikely destinations as Frankfurt and Hong Kong. United Airlines and TWA, once known for their lateral routes, invaded Florida. (In a typical TV commercial a little boy on an airplane says, "I've never been to Florida." Answers the TWA stewardess, "Neither have I.")

The scheduled airlines began calling at over 100 cities they had never served before. Big cities and big resorts were the gainers, but some cities, where the traffic was not as profitable as elsewhere, were suddenly stripped of service altogether. Commuter lines moved in, flying smaller planes and charging higher rates. The public was not always the winner. Allegheny Airlines, for example, later renamed USAir, pulled out of White Plains, New York, which served New York's rich Westchester County, studded with corporate headquarters of many large companies, among them Pepsico, Texaco, IBM, Reader's Digest, and, in adjoining Greenwich, Connecticut, the American Can Company. That terminated service to such centers as Syracuse and Washington, D.C. Both routes were picked up by commuter lines, but the fare to the nation's capital, which had been $100 round trip less than a year earlier, went up to $158, and even at that the business commuter had to reckon with small, uncomfortable planes. The safety record of commuter airlines, moreover, was less than admirable.

The new low promotional fares dangled by big scheduled airlines brought out a new class of traveler not conditioned to the long waits, frustrations and rules of airline travel. A strange by-product of deregulation was an air of fractiousness that broke out among passengers. Fistfights occurred in the aisles, causing United Airlines to issue a statement about company liability to its flight personnel. One Eastern Airlines pilot, unable to stop a fracas in the passenger cabin over smoking and nonsmoking seats, made good his warning and set his plane down at the next airport.

Passengers complained about the ever-shrinking knee room, the often interminable customs waits at Kennedy Airport and at Miami, the delays in check-ins and departures, the lines of passengers waiting in the aisles to use the toilets jockeying for space with cabin personnel pushing unwieldy meal carts. Flight attendants, harried and hustled, often turned sharp, even surly. It was a long way from the days of the smiling, always helpful, always cheerful hostess who could burp babies and fend off lustful traveling salesmen with equal finesse.

Despite discomforts, half of all passengers were flying on discount fares and many had a wide choice of airlines. Pan Am, after seeking the right for over thirty years, was allowed to fly across the United States. Moreover, it made use of its 747 SP, the initials standing for special performance. That meant the plane could fly from New York to Tokyo without a stop, or make it from Los Angeles clear to New Zealand in one overnight hop. While it did not provide berths on its long flights, as did Japan Air Lines and Philippine Airlines, it did install stretched-out seats that allowed the first-class passenger to recline in a sort of flying chaise longue.

With so many planes aloft, it might appear that ships, excepting Greek-owned oil tankers, might be consigned to Neptune's graveyard, to be written about only by aging chroniclers recalling a maudlin past of morning bouillon, afternoon tea, deck shuffleboard, gala nights, extraordinary cuisine and ladies arriving for gastronomic extravaganzas dressed to the molars.

It didn't happen that way. Although the *France* disappeared and the *QE2*, last of the ocean-crossing luxury ships, hung on the whim of its board of directors, a whole new era of cruising began. The maritime author Frank Braynard, the mind and motion behind the Bicentennial Year's Tall Ships extravaganza in New York Harbor, estimated that in 1979 there were more people buying ship tickets than ever before— except that they were now sailing on cruises. The demand for cruise ships was so great that many of the shipping companies had their existing cruise ships sawed in half and spliced with an addition midships, adding cabin space. By adding such a center section, a modest-sized liner could double its cabin space.

Moreover, the demand for cruise ships was responsible for bringing the *France* out of retirement. Sailing under Norwegian colors, it is cruising the Bahamas and the Caribbean with a new name, the *Norway*. Seattle entrepreneurs refitted the S.S. *United States*, still the holder of the Atlantic blue ribbon, preparing to sail her on long cruises for six months a year, and between the West Coast and the Hawaiian islands during the other half of the year. That crossing had been made regularly by the old *Lurline*, which used to arrive, with considerable panache, at Aloha Tower in Honolulu, while the Royal Hawaiian Band played syrupy melodies and well-wishers scurried aboard carrying flower leis to be draped around the necks of arriving friends.

During its half-year Hawaiian period, however, the *United States* is to maintain a triangular service from Los Angeles and San Francisco to Honolulu, then cruise for several days among the neighbor islands.

The owners adopted the time-sharing plan which has become popular among a number of resorts in the United States and abroad. Under this scheme a participant pays a flat fee which entitles him to use the apartment he buys for a specified time each year. Moreover, the apartment can be exchanged for similar use in a time-sharing resort elsewhere in the world.

Those who buy cabin space on the S.S. *United States* are entitled to cruise in that cabin for a specified time each year. In addition to the payment for the cabin they are charged 60 percent of the fare for the trip they elect to take. They also have the right to exchange their cruise time for a vacation at a dry-land, affiliated time-sharing resort.

There once was a time when more steamboats sailed the Mississippi than were in the entire British fleet. That number was now down to two, the old *Delta Queen* and the new *Mississippi Queen*. President Carter publicized cruising on the river when he took a midsummer sail from St. Paul to St. Louis in 1979. At Braynard's persuasion, the Soviet ship *Kazakhstan* made two three-day cruises of Long Island Sound in 1979 and sold out both trips. Moreover, the *Oriana*, taking advantage of the new popularity of New York, a circumstance brought about by the sinking dollar, arrived in Manhattan with 1800 passengers who slept aboard ship by night while touring the city by day.

Much of the credit for the revival in cruising was due to two factors: In a time of spiraling costs, whopping inflation in the United States and abroad, a vacationer could board a ship knowing exactly how much a trip was going to cost him. Secondly, the resounding success of the television show *The Love Boat*, featuring a triple-plot story line played out every week aboard a cruise ship of the Princess Line, had sent the clear message to young people, as well as old, that cruising was fun.

In the television version either the *Island Princess* or the *Pacific Princess*, its twin sister, is the "love boat" used for on-location shots. In these scenes the ships' actual passengers become the extras, their reward being to see themselves on television on some future winter's day when the snow is blowing. Other scenes are shot on two specially prepared soundstages on the Twentieth Century-Fox Hollywood lot. At the height of its popularity, *The Love Boat* was winning ratings in the 28 to 33 range while its competitors were making do with shows rating only 9 to 11. This meant that the happy-ending stories and the pleasures of cruising were reaching 30 to 40 million viewers in the United States. Even when it no longer wins prime time high ratings, *The Love Boat* is expected to be around, in syndication, for another ten years of reruns.

Officials at the Princess Line do not give all the credit for their success in winning nearly 100 percent bookings on every cruise to the show itself. They think *Love Boat* might have moved them from 90 percent occupancy to 100 percent a year ahead of time. The true popularity of cruising derives from an inflationary period when Americans, who had become experienced, freewheeling, around-the-world travelers, were suddenly faced with a shrinking dollar and ballooning prices. A cruise or cruise package combined with air fare to the port of embarkation provided travelers with a vacation bearing a fixed price tag. The only extras would be bar bills, shore excursions and tipping. There are those steamship psychologists who aver that the tensions of the late 1970s brought about by terrorism, political flash points, hijackings, kidnapings and inflation were contributing to the pleasant floating sanctuary and security that could be offered by a cruise ship, where one would be fed, entertained, romanced and then rocked to sleep like a babe.

What the shipping lines had done was to take the type of entertainment that had been so successful in Las Vegas and, in its heyday, in Miami Beach, and place it aboard ship. Specialized agencies such as the Bramson Entertainment Bureau in New York maintained a roster of two thousand performers—anybody from a bridge expert to a cruise director—who could be called upon to fill an entertainment slot on a cruise ship. Bramson handled the entertainment for more than a half dozen cruise lines. Working with highly developed computer systems, it booked magicians, dance teams, even clergymen. It could fill emergency defections in faraway ports, could place aboard any ship an automated audiovisual lecturer that would discourse (while showing pictures) on museums, archaeological digs or such tourist venues as Venice and the French Riviera.

The floating hotels of the cruise-ship world knew no boundaries. The fine ships of the Royal Viking Lines of Norwegian registry cruised the Russian Riviera along the Black Sea. The Carras Lines brought the first American tourists back to Cuba, along with a complement of American jazz musicians and enthusiasts. The Cunarders and the Holland-America Line were among the first ships to call at China. Docking at Whampoa Harbor for Canton or even at Shanghai became a commonplace excursion. Linblad sent a ship up the Yangtze.

Of the new places that had opened to America's irrepressible travelers, Cuba, after seventeen years of Castro, seemed a desolate place, served by new buses, officious chip-on-the-shoulder guides and aging hotels left over from what had been, for some, better days. The old Nacional Hotel could still stage a rousing show, but a buffet table spread along the line of cabañas that had known so many American visitors from other days was now, in the socialist society, marked with price tags. Some of the old bartenders were still on hand in the Floridita, which was once reported to serve the best frozen daiquiri in the world. The Floridita was the favorite of Ernest Hemingway, who came with his cronies from his *finca* outside town. Hemingway's villa is now a museum and the Cuban Government uses it as a tourist attraction, permitting visitors to walk on the terraces outside and look in the windows. Entering the house is forbidden, as is the taking of photographs. His old boat, the *Pilar*, named for a character in *For Whom the Bell Tolls*, is pulled up on the lawn, decaying.

Cuba is still able to produce a dazzling rhumba-belt show in the elaborate Tropicana where the stage is strung through the trees. But the shop windows are bare, and visitors are invited to buy in approved bazaars set aside for travelers. Cigars sold to foreign visitors are Cuba's prime tourist export.

On the other hand, the People's Republic of China, which by the end of the 1970s had opened wide for American tourists, proved for many the most fascinating trip of a lifetime. With the establishing of official relations between the two countries, visas could be obtained in Washington. Trips were arranged by specialists in the field, such as Lindblad Travel, and also by Pan American and Japan Air Lines. Regular tourists—as opposed to invited guests treated in VIP style—were arranged in groups of twenty-four led by a tour leader. This standard number made for manageable table service in hotels, similar-size buses and groups that were not so large as to become cumbersome when wandering through museums.

China Travel Service, the government travel agency, assigned two escorts from Peking, and a local guide met each tour as it arrived in a city. Standard tours began in Canton and ranged northward, exiting after several weeks from Peking. But there were other cities which the persistent could arrange to see—Loyang, Sian, Lanchow and Urumchi. Tibet opened in the spring of 1980.

Travel to China became a real status symbol among Americans. It was no longer enough to say that one had just come back from a trip around the world. Indeed, if two veteran travelers met, one might open with a gambit like this:

NO. 1: Meg and I have just come back from China. Fabulous trip.

NO. 2: Oh, was this your first visit? [elegant put-down]

NO 1: Yes, and we brought back some beautiful jade.

NO. 2: We thought our second trip was more interesting than our first. How far north did you get? [setup for a put-down]

NO. 1: We went all the way from Canton to Peking. Covered two thousand miles. Saw everything.

NO. 2: Really? They let us go to Inner Mongolia. You should see the fur hats we bought.

Not even the spartan and sometimes downright seedy hotels seemed to deter the American fascination for Cathay. They came with cameras, sound movie equipment, tape recorders and traveler's checks. They sat through endless discourses on how communes are operated, how factories are run, how kindergartens are managed. Some got inside air-raid shelters. Everybody got a look at the Forbidden City and the splendor of the Summer Palace. Almost everybody got to eat a Peking duck in Peking, where—as the word filtered down—there were three main Peking duck restaurants to choose from. The largest of them was called Big Duck, the smallest, Little Duck. The one near a hospital which the Rockefellers had subsidized, inevitably became known as Sick Duck.

The opening of parts of China to American travelers in the late 1970s, and the promise of unveiling other parts of the country which had been unseen by the Occidental eye for decades brought a heightened interest in travel to Asia. China proved more than a compensation for the tragic loss of Cambodia, whose temples at Angkor Wat had been a prized destination in the years before the Vietnam War spread across the borders and Cambodia was wracked by the overthrow of Prince Sihanouk, the overthrow of the Lon Nol government by the fierce Marxists, and finally, the invasion of the country by the Vietnamese and the massive starvation that ensued.

Nonetheless, travelers with a Chinese visa in their passports were including Japan in their itineraries (despite the enormous rise in the cost of tourist life for dollar-carrying travelers), as well as Hong Kong (where rooms were always at a premium), Singapore, Bali, Java, Thailand, the Philippines and Taiwan.

The string of American hotels—Hilton, Hyatt, Sheraton, Inter-Continental, Western, Marriott, Regent and others—was stretching yet deeper into remote areas of the world. To the traveler far from home the promise of a Hilton-operated game lodge in Tsavo Park in the middle of Kenya was an assurance of comfort just like home. A Nile River cruise was exotic travel in supreme luxury.

American-style hotelkeeping took on a special character, a certain flamboyance that perhaps can be traced to the opening of the Hyatt Regency in the middle of downtown Atlanta in 1967. A creation of John Portman's, it sported lighted glass elevators sliding up and down the walls, a huge atrium and an enormous impressionistic flower rising twelve stories out of the lobby floor. An architectural tour de force, the Atlanta hotel became the forerunner of a whole chain of Hyatts noted for their daring design. It also marked the emergence of Atlanta, once a drowsy town, as one of the supercities of the South, a metropolis where lighted elevators sliding up walls were as much a necessity as a row of bellmen.

There was so much going on in the lobby of the Peachtree Plaza Hotel, unveiled in the mid-1970s, that some compared it to a set for *Star Trek*, the popular TV series about outer space. Fountains burbled, birds twittered, water cascaded over falls or sat limpid in pools. And, of course, elevators shot skyward, two of them through glass tubes—a dual express to the seventieth floor, where a tower restaurant turning slowly offered an ever-changing view of the city.

The Peachtree Plaza Hotel was followed by the Omni International, part of a center that included office buildings, a convention hall, six theaters, a sports arena and an ice-skating rink. Omni went on to construct a similar extravaganza in Miami that became a wondrous place to visit, not merely a hotel.

Starting in 1971 with a western hotel called Houston Post Oaks, which had everything from a skating rink to branches of Neiman-Marcus and Sakowitz Bros., Houston—at least, downtown Houston—became an architectural exercise with buildings designed by I. M. Pei, Philip Johnson and other leading architects. Hyatt put together a cavernous hotel at Embarcadero Center in San Francisco, a building with a lobby so large that football players were brought in to practice punts as a publicity exercise. Embarcadero Center became a *place*—some called it Rockefeller Center West—a touristland with fifty shops and restaurants covering eight and a half acres that had once been a seedy waterfront slum.

The emergence once more of downtown as a place to go proved no less dramatic in Los Angeles, where a billion dollars was spent. As New York became the Big Apple, L.A. became the Big Orange, creating a stunning assortment of theaters and music halls—all downtown. Western International imported the Atlanta impresario-architect John Portman, who produced five cylindrical glass towers collectively called the Bonaventure, a hotel with 1500 pie-shaped rooms reached, of course, by lighted glass elevators sliding up walls. The effect caused the Los Angeles Biltmore, a neighbor built in 1923, to indulge in a massive face-lift that has returned it to its flapper-era opulence.

Nowhere was the national sense of a return to downtown more evident than in New York, which had always been the tourist's city. But during the early years of the 1970s, Gotham had seemed to fall on hard times. Large companies were leaving, bankruptcy loomed and crime, fanned by the spread of the drug culture, gave the city an unsavory reputation. Central Park was said to be a no-man's-land where muggers lurked. Side streets were alleyways of danger.

But then, as the decade drew to an end, the mood seemed to change. If there was a turning point perhaps one could identify it as that splendid day July 4, 1976, at the very epicenter of Bicentennial Year, when the Tall Ships paraded in New York Harbor and the whole city danced, celebrated and lit up the night with fireworks. From then on New York began to come back. Two publicity campaigns—the coining of the Big Apple slogan, and the I Love New York promotion, together with the devaluation of the dollar—made New York—for Europeans and Japanese—a new inexpensive wonder of the world. Visitors to the city tripled in the three years before the decade ended and there were twice as many visitors in town from overseas as there had been ten years before.

Among the new attractions was the great World Trade Center with its skytop restaurant, Windows on the World, offering incredible views of the city and the harbor, and some unusual cuisine to boot. On the ground floor, the Market Square Dining Rooms and Bar were a midweek dining hall of special note, but on Sundays they were a happening with bands playing, greengrocers offering their wares, and in the fall, a celebration of apples in what was, after all, the Big Apple. You could even buy a poster to take home to prove it.

Uptown, Citicorp Center, an arresting architectural exercise intended as a place of business, had become a tourist attraction on its own. Visitors flooded the atrium with its circle of restaurants and shops. They joined New Yorkers to watch the free entertainment presented in the atrium during the after-work cocktail hour.

New hotels began to sprout. Hilton International placed the first of its domestic Vista hotels alongside the World Trade Center near the southern tip of Manhattan. The new Palace, an elegant establishment, rose in midtown between Madison and Park and Hyatt had all but rebuilt the old Commodore Hotel next to Grand Central Terminal. Aside from the great chains, America was adding to its string of small and refined hotels. Such vaunted establishments as the Stanford Court, a make-over just off the side of Nob Hill in San Francisco, where fancy cars could drive in and park under a glass dome, was a classic success. Inside there were television sets to shave by, Baccarat chandeliers that had once hung in the Grand Hotel in Paris, marble floors quarried in Carrara and a restaurant called Fournou's Ovens lined with private wine bins of faithful clients and equipped with a 35-square-foot oak-burning oven.

In Washington the Sheraton people undertook a major face-lift of the Carlton on K Street which, inspired by the Meurice in Paris, had opened in 1926. The Fairfax, completely redecorated by John Warnecke, quickly achieved a reputation for supreme elegance. Handsome and beautifully managed, both could vie with the attractive Madison and Loews L'Enfant Plaza, where the telephone operators answer with a sprightly Gallic *Bonjour.*

All Americans had known Luna Park and Coney Island. Amusement parks were part of the American scene as they had been in Europe. Tivoli in Copenhagen had been the progenitor of playlands the world over, and its very name had virtually become generic. But when Walt Disney, the father of Mickey Mouse et al., opened Disneyland in California, it marked a departure in amusement parks. They were to become known as "theme parks," and they became a large part of American leisure time.

Disneyland, near Anaheim, was imaginative, clean and run by a staff of pleasant people. And while it made a great splash when it first opened outside Los Angeles, it was only a warm-up for Walt's biggest venture, Walt Disney World, near Orlando, Florida. Here the Disney people bought forty-three square miles, a parcel of land twice the size of Manhattan Island and at least 250 times larger than the Vatican.

The master plan called for an amusement park to be called Magic Kingdom, but it went far beyond that. Disney World included half a dozen hotels, condominium apartments, golf courses, campgrounds and an office park. It kept automobiles on the perimeter and required residents and visitors to make use of a transport system of monorails, boats and railroads as well as golf carts and bicycles for short journeys. Three miles of waterway provided plenty of room for steamboats, or, in the "wilderness," a stream in which to paddle a canoe. Paddlewheelers and steam launches slipped along the waters. Restaurants grew along the shores. So did a whole water playland as well as a dry-land ranch, campsites and a hotel with 1100 rooms that was so modern that a monorail ran right through it. At the same time, the Florida version repeated the California success by creating again "Main Street," America as it might have existed at the century's turn with horse-drawn trolley cars, early automobiles and double-decker buses. In the Haunted House, an instant hit, disembodied heads talked, headstones shook back and forth and ghosts pirouetted across banquet halls. There were also some attractions of parochial interest—the Pirates of the Caribbean, for instance, an elaborate show in which visitors traveled in small craft watching the picaroons ravage and plunder.

Besides all the playfulness, Disney had more serious intentions for its Florida tract, among them a center where new technologies in energy, transport, agriculture, health and medicine, space and oceanography will be researched and tested. A World Showcase, or, so to speak, a miniature world's fair on permanent view, is planned with nations building cafés, theaters and shops along with industrial displays, all of which will be manned by a corps of young people from various parts of the world.

While the lofty intentions were left largely to Disney, the success of the California Disneyland and of Disney World inspired a string of parks all over America, some of them underwritten by large companies such as Marriott and Anheuser-Busch. The St. Louis-based beer company had parks at Tampa, Florida, and Williamsburg, Virginia.

Aside from visits to manufactured parks, there were those, perhaps inspired by the new American appreciation for the environment, perhaps driven by the pressures of urban stress, perhaps even encouraged by the new leisure, who sought a different kind of holiday in the out-of-doors. The national parks, especially those nearest centers of greatest population, drew unprecedented numbers of visitors, often causing serious overcrowding which contradicted the original reasons for which the parks had been set aside.

The glut of roisterers, trippers and campers leaving behind empty beer cans and bad memories became a real incursion into what had been intended as pristine wilderness preserves. Nearly 12 million visitors a year were pouring into Great Smokies National Park, over 15 million were touring the Blue Ridge Mountain Parkway and 2 to 3 million were being clocked in Yellowstone and Yosemite. Where once there were bear jams caused by bruins panhandling along the entrance roads to Yellowstone, there now were people jams. Brightly colored inflated rafts were floating down the streams. Rock music blasting from portable radios and stereo sets split the stillness.

To divert some of the traffic the National Park Service published a book that was both plea and suggestion. It was titled *Visit a Lesser-Used Park*, and it emphasized that the federal park system also included national recreation areas and national seashores, as well as historical and commemorative sites. There was plenty of room, for example, in Utah's huge parks in a corner of the country where interstate highways do not cross. Safari wagon trips and river rafting excursions were popular in Utah's Canyonlands, and visits to Arches National Park, with its odd rock formations, provided a new sense of America. There were natural wonders, too, at Craters of the Moon National Monument in Idaho, where the earth's crust hissed and bubbled.

Those seeking a sense of history in the outdoors could look to Antietam, where Lee first invaded the North in 1862, and Appomattox, where he surrendered three years later. Both were equipped with picnic and camping grounds. Limitless wide-open spaces surrounded Little Big Horn, where Custer heard the hoofbeats of the Sioux and the Cheyenne.

Shunpikers left the beaten track to find Indian ruins at Chaco Canyon in New Mexico; unfettered Gulf islands a boat ride from Biloxi, Mississippi; and brown bears and bald eagles in the Valley of Ten Thousand Smokes at Katmai, Alaska.

Resorts, American and foreign, seemed to have a certain lifespan dictated by cycles, moods, public fancy and such very real influences as transportation and weather. Miami Beach, which boomed in the 1930s, and again in the decades following World War II, began to decline when the airplane began lifting people at affordable prices to Caribbean and Mexican destinations where the winter weather was certain, the beaches were long and sandy and there were even elements that were faintly exotic.

Atlantic City, in the 1920s and the 1930s a nearby resort for New Yorkers, Philadelphians and New Jerseyites, never really recovered after World War II. Once a showplace of huge hotels, a long boardwalk and elegant shops, it was a seaside hideaway where playwrights and composers could try out shows, a place to bundle up in the colder seasons in a deck chair and steamer rug to come home with a facial tan.

But in spite of Miss America beauty pageants that were a part of every September, Atlantic City began to crumble, to sink back into the sea that had spawned it. Only one thing, it seemed, could save it: legalized gambling. In November 1976, the voters of New Jersey, perhaps riding on the euphoria of Bicentennial Year, voted to approve gambling. The town went on an all-night toot. Property values zoomed, dreams seemed suddenly to become reality.

In short order Atlantic City began to transform itself. Before the decade ended, a month more than three years after gambling became legal, there were two casinos running at all hours, with waiting lines at many gambling tables. Resorts International, a pioneer in the renaissance of Atlantic City, was raking in $800,000 a day at its casino on the boardwalk at North Carolina Avenue. On the premises were 1354 slot machines and 129 tables for an assortment of games of chance from roulette to craps. Caesar's Palace, a landmark in Las Vegas, opened the Boardwalk Regency Hotel, complete with casino as in Nevada.

There seemed no reason to attract customers; they were coming anyway. But the momentum was rolling and so Atlantic City began to book big-name entertainment. Las Vegas–Folies Bergère-type revues with leggy showgirls became a fixture, and so did lounge entertainment, a sort of sideshow to the big names and to the main draw—the gaming tables.

The old hotels were demolished, one by one, but all sorts of plans for new lodgings were on drawing boards, including a venture by no less respectable a company than Hilton International.

The year that Atlantic City got its gaming license would be remembered for that treasured piece of permissive legislation by croupiers, gambling resort operators and all the citizens of the seaside salt water taffy city. But the rest of the nation would remember 1976—especially July 4—as the great party to celebrate the two hundredth anniversary of the United States of America. It was a time for reflection, for a wave of terrorism was already shaking the world.

Meanwhile, in New York, some 52 warships from 22 nations arrived for the greatest of the Bicentennial celebrations that swept the United States. They joined the tall-masted sailing ships—more than 225 of them—that had gathered in lower New York Harbor for the parade up the Hudson. President Ford was flown to the huge aircraft carrier *Forrestal* to take the review, then was lifted upriver to the U.S.S. *Nashville*. But the President was also at Valley Forge that day, where he stood on a covered wagon that represented Michigan, his home state, and signed a bill making Valley Forge, where Washington's colonials had spent so bitter a winter, a national historic site. From Valley Forge he went to Philadelphia, the city with perhaps the deepest American roots, where a million people greeted him. The Liberty Bell was struck with a rubber mallet, a scene that was carried by national television all over the country.

July 4, 1976, was a day that had started in the United States on Mars Hill Mountain in Maine, where dawn first ignited the continent, moved on to Fort McHenry in Baltimore Harbor and ended in a huge fireworks display that night in New York Harbor. In Rome, the Pope blessed the United States. In Japan, citizens visited U.S. military bases. In Mexico City, a party was given for 607 American prison inmates, most of them there on drug charges. In Wooster, Ohio, a restaurateur made a five-thousand-pound blueberry sundae, added whipped cream and American flags and placed it on a thirteen-foot-long platform. Of the national birthday party, McCandlish Philips wrote in the New York *Times*, "It was an exercise in percussion, procession, demonstration, declamation, detonation, commemoration, vociferation, trivialization, solemnization, and, for some, indigestion." But it was a year and a day to remember.

The publication of Alex Haley's *Roots*, which became a national bestseller, and the television series which followed it during the winter of 1976–1977, touched off a new American interest in finding out where one came from. Most blacks might have more problems than Haley in tracing their ancestors to Gambia or other parts of West Africa, but the chase was hot in Britain and in Ireland, where genealogical offices were swamped. Travel agents offered "Roots" tours and Americans of whatever European ancestry joined in the homecoming touched off by this book written by a black American. Pan American turned to television, radio and print to announce, "Every American has two heritages. Let Pan American help you discover the other one." The only complaint came from the American Indians, who said, "We have only one heritage; we don't have to go anywhere."

A new style of travel, rather like a hyperenergized adult camp, crossed the ocean from France, where it was born, and opened branches that would attract the American market. The project was called Club Mediterranee—in popular parlance, Club Med—and the approach was all new, a little Gallic with a touch of South Pacific Tahitian here and there. For one flat price, the club offered a full week or two at one of its villages. Included were transportation, lodging, splendid and bountiful meals, wine, sports and endless entertainment. The "counselors" were called *gentils organisateurs*—GOs for short, and the enrollees were called *gentils membres*—GMs for short.

A grand success in such diverse locales as Moorea in French Oceania and Marrakesh in Morocco, Club Med moved into Mexico with resorts at Cancun and Playa Blanca, into the French islands of Guadeloupe and Martinique in the West Indies, into the Bahamas at Paradise Island and Eleuthera, into Haiti, and at last into the middle of the United States itself, at Copper Mountain, Colorado.

Internationally, the pressures of the Middle East political pot, that bubbled continuously, and the creation of OPEC, a cabal of oil-producing nations, had their effect on travel and on world economy. An embargo in the winter of 1973 caused lines to form at service stations which threatened such drive-in vacationlands as Disney World in Orlando, Florida and Colonial Williamsburg in Virginia. More serious was the fall of the Shah in Iran and the collapse of that nation into political shambles. Gas lines formed once more in the summer of 1979, seriously affecting travel by car in the United States, and from the United States to Canada and Mexico. The price of gasoline more than doubled during those six years, and although it was still well below the cost in some European countries, the specter of spending $20 or more to fill the gas tank of the family car turned people away from the highways. (Contrarily, both Mexico and Canada could offer bargains at the gas pump and the advantages of touring in those countries were not overlooked.)

Amtrak, the National Railroad Passenger Corporation, found a new popularity. The line, which had brought the nation a semblance of decent railroading, was about to lose considerable service when the oil crisis hit, and people turned to mass transit. Suddenly berths were sold out on the more popular routes, and on the regular service along the northeast corridor it was often impossible to find a seat. It made little sense to curtail mass transportation in the teeth of a fuel emergency. Transportation eats 30 percent of all the energy consumed in the nation, the automobile taking more than half of all the petroleum used in transportation, not including that required for building the cars in the first place, or for construction of highways. The train could move more passengers more miles on a gallon of petroleum-based fuel than any other form of transport.

At decade's end Amtrak found itself trying to hold on to its routes and equipment while at the same time promoting its new fleet of deluxe Superliner coaches, which had two decks connected by a stairway. Sleeping cars had deluxe and economy-class bedrooms; the diners, a full-length upper-level dining room. Piano bars enlivened the mood in the lounge cars. There were full-width bedrooms for families as well as special accommodations for the handicapped.

Nearly three hundred cars were ordered, most of them to be assigned to long-distance Western runs and to transcontinental routes. Seeing the United States of America by train loomed as the new old treat. But even with the new equipment, Amtrak would still not be meeting the style of European railroads, all of which, except for a handful in Switzerland, are nationalized. In the countries to which Americans travel most—that is, Britain, France, Germany and Italy—new rail service will connect cities on trains that fly over the rails at nearly two hundred miles an hour. French railroads have offered vacation cars that carry pinball machines, jukeboxes and even lectures on travel. Special rail equipment accommodates bicycles and horses for those riders who want to go by train one way only. Autos can be put on freight cars while the driver sleeps away the night in a berth. Despite all those attractive improvements on the rails, the railroads in Europe often suffered the same problem as those in the United States: no porters, a worldwide shortage that has caused luggage manufacturers to put wheels on baggage.

Over two hundred years ago Americans had started traveling between the colonies by stagecoach. The natural descendant of the horse-pulled stage is the bus. By the end of the 1970s there were over 1000 bus companies traveling between cities. Their buses were pulling into stations or stopping along the curb at 15,000 communities. All but 1000 of these settlements depended upon the bus as their only public-transportation link with neighboring towns. The buses covered nearly 300,000 miles of routes. The sleek coaches of Greyhound—some 4500 of them—accounted for more than half the business. The number two bus company, Continental Trailways, carried less than half as many passengers as Greyhound on half the number of buses.

It is a curious fact that during the height of World War II, with gasoline rationed, seats on planes hard to find and many servicemen and families trying to get from here to there, buses were carrying half a billion people a year. They have never done that kind of business since, although they rose to about 351 million in 1979, much of that increase caused by the high price of gasoline. But that number, though not nearly as high as the wartime peak, is still more people than are carried by Amtrak and all the airlines put together. After two centuries of travel, and a nation expanded to fill the continent from ocean to ocean, for many citizens the gas-powered stagecoach was still the way to go.

As the 1980s unfolded some saw a whole new social structure beginning to emerge. The pattern of the old days of the long grand tour complete with steamer trunk, which had given way to hop-skip-and-jump trips with suitcases through a maze of countries, was altering yet again. People seemed eager to get away—away from inflation, terrorism, politics, even from the household itself. And they needed to do it several times during the year. It was the ultimate restorative.

"People want to get away even if it is just for one night," reported Joseph Kordsmeier, the senior marketing expert for

Hyatt's hotels. He spotted a social change taking place in the first months of the new decade. All over the nation and in Canada hotels were organizing get-away-from-it-all weekends. Frills were added to sprinkle glitter on the idea—breakfast in bed, bottles of champagne, a dinner for two in the hotel's premier dining salon.

Perhaps the epitome of weekend hopping was reached by Kordsmeier's Hyatt Hotel in Honolulu, which cooked up a so-called honeymoon package for young Japanese. The couples flew through the night from Tokyo on Japan Air Lines, arriving in Honolulu at 8 A.M. They had all day to spend on Waikiki Beach with time for shopping along Kalakaua Avenue. For ninety-five dollars per couple they were given a room for a night at the towering Hyatt Regency across the street from the world-famous beach, plus dinner for two, and continental breakfast the next morning. Then JAL flew them back to Tokyo. As hectic as that schedule seemed, fifty to seventy-five Japanese couples a day were taking advantage of the package.

Getting away from it all had become a necessity. More than other forms of temporal escape, travel could lift one from the treadmill of dailysville, put one down somewhere else, and provide a whole new aspect for the bruised and battered psyche.

X | *The Spaced-Out Traveler*

WILL WE ONE DAY really travel in space, visit the moon, journey to space colonies, or is that all *Star Trek* science fiction? The gospel according to Isaac Asimov, the seer of science writers—if you are willing to accept his 20/20 foresight—says that those who are the young people of today may well have a chance to rocket off to the far blue yonder. As for their grandchildren, such excursions will be as routine as catching the next jet to Chicago is today.

To accept his theory, Asimov urges us to remember that the first moon walk took place in 1969. If you put yourself back a hundred years before that you would be in 1869, still clucking over that amazing feat, the laying of the Atlantic cable. You would be without electric light, which was not to be discovered for another ten years. Automobiles were two decades away and airplanes would not begin to fly for another thirty-five years. Moving as fast as we are now, however, there is no question, at least in Asimov's mind, that a hundred years hence anybody who can afford a ticket and get a reservation will be a space traveler.

Writing in *Adventure Travel* magazine just before the 1970s became history, Asimov predicted huge space colonies traveling in the moon's orbit, as well as an assortment of power stations, factories and laboratories all floating in space. As for the moon itself, it will ultimately be made into a world much like the earth. That might take more than a hundred years, but even on a moon without water or air, tourists will travel over the lunar surface in enclosed vehicles to see the mines and the factories. "There will undoubtedly be elaborate hotels under the lunar surface," Asimov writes, "where visitors can be completely comfortable in a thoroughly simulated Earth environment." Weightlessness will just add to the fun, he says. "The time may come when 'Get yourself in shape on the Moon' will be the slogan of the Beautiful People of the twenty-first century."

What will impart a real ring of probability to Asimov's fancies will be the comings and goings in the 1980s of the space shuttle. Since it has wings and will return, more or less conventionally, by making a landing at an airport, the space shuttle will be blessed with public recognition. Its first mission into space, portentous as it may be, will not be the public's first look at this strange craft. It appeared in September of Bicentennial Year bearing the name *Enterprise*, which happened to be the name of the *Star Trek* spaceship of Captain Kirk and Mr. Spock. The word figured, too, in John Kennedy's speech of exhortation in May of 1961, when he called for the landing of a man on the moon. "Now is the time to take longer strides—time for a great new American enterprise—time for this nation to take a clearly leading role in space achievement, which in many ways may hold the key to our future on Earth. . . ." So went President Kennedy's proclamation.

The same Dr. Asimov, writing the foreword to a book called *Enterprise* by Dr. Jerry Grey, about the shuttle and our future in space, argued that our exploits in space, culminating in the moon landings, could be written off as highly expensive adventures, as stunts. When the game was over, he noted, and the race with the Soviet Union won, the public lost interest in "space spectaculars."

But this means abandoning space flight in its very childhood. "Space flight comes of age," Asimov wrote in Grey's book, "when it becomes cheap enough to do something more than stunts. When it can make homes for human beings in space. When it can be used to draw on resources of energy and material from space for the good of humanity. When it can be used to extend human knowledge of the universe and human ability to manipulate and make use of the universe."

Once we have solar-power stations and lunar mining stations, the cost of getting to them and returning to earth will not be a detracting expense. Politics and publicity won't energize space technology; profit, straight economics and the chance of a better life will be the motivating forces. "The real beginning of the space age," Asimov said, "is right now."

A scenario for the future appended to Grey's book predicts an exciting third century for America with a satellite delivering solar power back to earth by 1992, fusion-powered spacecraft leaving for Mars by 2012; an interstellar vehicle taking off for Barnard's Star by 2062 with 1000 carefully chosen colonists on board. And then on July, 4, 2076, the third American century, a landing on an extrastellar planet by people from earth.

That is all mighty heady stuff, today's traveler might say, but what about next year's vacation? What will I be able to do five years hence that I'm not doing now?

Aeronautically, very little. The 1980s will be the first decade since the end of World War II not likely to be marked by revolutionary development comparable to the successive arrival of jet travel at the end of the 1950s, the emergence of the wide-bodied airplane in the 1960s and the appearance, even in limited fashion, of the SST in the 1970s.

The fuel crises that marked the 1970s have motivated airlines and airplane manufacturers to seek lighter planes that are more fuel-efficient. Boeing will continue to produce the 727, which has proven one of the most popular airplanes in the skies, but that craft has been in production for fifteen years and Boeing must redesign its wings in order to make it more fuel-efficient. At the same time it will be introducing the 767, a transcontinental airplane which will be larger than the 727, but smaller than the jumbos. Its 767 and the new 757 are designed to be 30 percent more fuel efficient than the planes they are destined to replace. The 747, first of the jumbos, will be "stretched" in two ways. The upper deck, used as a lounge by some airlines and as a first-class section by others, will be lengthened. So will the midsection of the aircraft. By mid-decade the 747 will be able to carry 500 to

600 passengers, 100 to 200 more than are now transported at full capacity. Similar stretches are planned by McDonnell Douglas for its DC-10. McDonnell Douglas has a DC-9/80 coming which will have a better fuel efficiency than the original DC-9, but which may have to compete with Europe's Fokker F-29. By mid-decade the European-manufactured Airbus A 310, wide-bodied, fuel-efficient and low on noise, will be carrying 210 passengers on short- and middle-distance routes.

The nagging worry over fuel efficiency has given a lift to Airbus, and also to Lockheed, whose L-1011 rates high in that department—forty-six passenger miles on a gallon of kerosene, just one digit better than Airbus. Nonetheless, Boeing was able to beat out the new Airbus and land a huge order from TWA for the 767.

So determining is the cost of fuel to airlines that Pan American, deciding that Lockheed's L-1011-500 was the most fuel efficient airplane yet produced, ordered a dozen and took options for fourteen more. The first of them went into service on Pan Am's Latin American routes, running along the East Coast to South America. What made the difference for Pan Am was a four-and-one-half-foot extension on each wing tip and new Rolls Royce engines making the Dash 500, as it is called, 22 percent more fuel efficient than Boeing's SP. First-class passengers are seated six abreast, in units of two, with two aisles. Economy-class seating has three seats on either side of the plane and four abreast in the center. For Pan Am, which owns forty-five Boeing 747s, pioneered the Boeing 707 and the Boeing Stratocruiser, as well as the great B-314 flying boats that plied the Atlantic and the Pacific, the switch to Lockheed was a significant step. The airline had not flown Lockheed airplanes since the days of the Constellations. Now the new economics of oil had dictated a billion-dollar contract.

The crystal-ball gazers at Stanford Research Institute fore-see the zenith of gasoline production being reached in the early 1980s and diminishing after that. Planes, cars and every-thing else that runs on petroleum products will have to seek new ways to get better mileage or else find new propellants. Lockheed makes much of a hydrogen-powered aircraft. Its executives see shortages in airplane fuel early in the 1980s but nobody doing much about it until decade's end. Lockheed says it could convert present carriers to hydrogen fuel, but in the best of all worlds it would like to build a hydrogen plane which would look like a large, winged thermos bottle with seats fore and aft of the fuel container. The fuel could actually be manufactured at airport sites. Lockheed has sketched out sixteen such places which would account for the bulk of the world airplane traffic. A water source and a power source— coal, nuclear, solar or whatever would be enough to create liquid hydrogen, which could be stored in tanks near airfields.

A fanciful idea somewhat more distant in the future—perhaps thirty years away—is a flying wing that could carry 1700 passengers as well as 650,000 pounds of freight. H. W. Withington, engineering vice-president of Boeing, calls the wing "a hypothetical vehicle designed to give passengers the lowest-cost fare over a long-range trip." A plane that weighs a million pounds, which is three times as heavy as a 747, would need a huge wing to lift its load. Since its payload capacity would be far larger than needed just for passengers, it could carry freight, too. It would be limited to use between large urban airports.

Boeing's Withington also postulates the possibility of nuclear-powered aircraft, but the weight of the reactor and its shield would make its use feasible only in enormous airplanes of a million pounds gross weight. Reactor weight remains the chief problem, for it has already been proven by newly designed reactor shields that a reactor could be contained in the event of a crash. A nuclear-powered airplane of conventional design could fly for ten thousand hours without refueling.

Where does all this leave the supersonic transport for which there were such hopes (and such opposition) in the 1970s? Stanford Research Institute, one of whose clients is the Boeing Company, doesn't see more than ten faster-than-sound airplanes flying during the 1980s. That estimate was made despite the surprising fact that on the North Atlantic run, as the 1970s drew to a close, Concorde, the British-French SST, was gaining a coterie of adherents and was making money. By the 1990s, however, it is anticipated that a much larger supersonic transport, with a titanium body, such as that originally proposed in the United States' SST model, will be built both by American companies and by some foreign consortium tying together European and perhaps Japanese capital. The competition in airplane industries on both sides of the Atlantic is keen and both Europe and the United States need to maintain their employment levels, to say nothing of their respective balance of payments, which rest heavily on the airframe industry.

NASA has funded a research project for a hypersonic airplane which could cruise at Mach 6, that is, six times the speed of sound at sea level, or four thousand miles an hour. The design concept being developed by Lockheed-California calls for a dual propulsion system using both conventional turbojet engines and supersonic combustion ram jet engines (called SCRAM), fueled by liquid hydrogen. The turbojets would power the aircraft on takeoff and on landing and be used as well to bring it up to the speed at which the SCRAM jets would take over. The closest analogy would be the old gear-shift system in a car, in which low gears are used to bring the vehicle to normal cruising speed. Including subsonic climb and subsonic landing to stay with noise regulations, the flight from Los Angeles to Tokyo, carrying 200 passengers, would be 2 hours, 18 minutes.

Many of the breakthroughs in aviation will not necessarily be as visible to the traveler as tomorrow's proposed hypersonic plane. What is emerging in the immediate future is an automated age in which computers, once activated, can arrange for takeoff, piloting and landing of large aircraft. Devices already extant notify the pilot of approaching difficulties, some of them even capable of delivering immediate voice commands, eliminating precious seconds in thinking time.

But that is only the beginning. Computers will revolutionize the automobile, change the back-of-the-house methods of hotelkeeping, and make possible high-speed trains that will move too fast to be controlled by human minds and hands.

Microprocessors, in reality silicon chips etched with a myriad of electronic components, built into the family car, will be able to take over braking an automobile traveling over potentially dangerous surfaces. Where the driver might be inclined to slam on the brakes at the start of a skid, the computer will cancel out the driver's foot action, and activate the brake so that it will be "pumped" in measured fashion and be less likely to skid, or "hydroplane." Computers will warn the driver of imminent danger—a car approaching in the wrong lane, a slick surface around the bend, poor visibility ahead. The microprocessor also has the capability to compute the most advantageous speeds for fuel consumption. Ready for installation now, but awaiting legalities, is a drunk-driver interlock, a system that would require the driver to match a moving needle on the dashboard with a second needle that moves under manual control. If coordination has been affected by drink the car will not start. Among the other options are computers that will calculate how many miles can be traveled on the gas remaining in the tank. Maps that come in cassettes will be displayed on a screen, along with the best routes through traffic, a signal sent by local police headquarters.

As recently as 1975 such a complex computer system was thought to be too wildly futuristic even to discuss. Pushed by the government's flat dictate to bring average fuel consumption to 27.5 miles a gallon by 1985, automobile manufacturers forced their designers and engineers to come up with new concepts, however startling. One was the adaptation of space techniques to automobiles. Another was the move to build lighter cars out of heavy-duty plastic rather than metal. Although personal risk might be increased by the substitution of plastic for metal, the new computer safety techniques will, it is hoped, offset those potential dangers. Cars will be made to use two kinds of energy at the discretion of the driver: battery power for low speeds such as driving from house to supermarket, and gasoline for long hauls.

It has been said that computerization of the car will be like having a mechanic under the hood. One executive of the Ford Motor Company believes that the impact of computers on the public "will be bigger than automatic transmission," which did away with the gear shift and the clutch.

Pressed for a more efficient use of fuel to move people, the European railroads mounted a vast improvement program to speed up systems and attract riders. France will be spending $1 billion to produce fast trains that are expected to attract 25 million passengers a year, many of them wooed from the nation's highways. High-speed lines running to the southeast from Paris will cut two hours off the route to Marseille. In Italy, straightening the highly traveled tracks between points such as Rome and Florence will cut that run down to 1½ hours. The high-speed trains will eliminate those meals in the grand fashion which were an elaborate (and expensive) ritual on the European railroads. Now the tendency is to use half of one car for snack service in the fashion introduced by Amtrak in America.

The most dramatic departure in railroad travel will come from Germany, where Krauss Maffei, an industrial company, together with Messerschmitt and Thyssen, the steel company, are in the last stages of producing an ultra-high-speed train that will operate by means of magnetic levitation. Federally funded in 1969, the train, now called Transrapid, has undergone a number of mutations since the first model appeared in 1971. A rocket sled that appeared subsequently got good speed but the hardware didn't hold up, and the noise and pollution were inadmissible.

Magnetic levitation, or MAGLEV, operates on the principle of magnetic rejection. Similar poles—one on the undercarriage of the train, the other on the track—repel each other, causing the train to rise and to be propelled on a cushion of air. Edges of the carriages are curved around the track for guidance. By 1985 to 1988, Transrapid is expected to be in service with two-car trains carrying a total capacity of 200 passengers, operating between cities at a speed of 250 miles an hour. The track, which can be placed along existing roadways either elevated or at ground level, would connect such cities as Boston and New York, Montreal and Toronto, Los Angeles and San Diego, relieving those centers of congested air traffic as well as congestion on roads connecting city core with airports.

There seems less that is revolutionary on the sea-lanes

in the future than in any other form of transport. As with other systems of locomotion there has been a rush to seek alternative fuel systems. Hydrogen and nuclear systems seem a long way off, but naval engineers are hoping to learn lessons from the Navy's successful use of nuclear submarines and nuclear surface ships.

As passenger ships became less a mode of transportation and more a floating resort, there was an effort to standardize staterooms, which would ease marketing and sales. To that end there has been experimentation with a catamaran or twin-hull design which would permit a tall, squared-off hotel structure to be placed on a large platform. The shallow draft would allow greater speeds and the rectangular platform would allow a more efficient hotel structure.

The return of the *France* as the *Norway* sailing out of Miami for the Caribbean is a display of the potent future in short-term pleasure cruising. The reappearance of the S.S. *United States*, the holder of the Atlantic blue ribbon in another era, means that yet another former liner will become a floating hotel.

A faint but unmistakably starlit glimmer shines in the eyes of some Danish shipping people who think cruises have turned too sharply toward the mass market. A division of the Danish Lauritzen shipping interests is planning a luxurious ship that by 1982 will be plying between New York and Freeport in the Bahamas, carrying 400 cars and 1000 passengers. Among its trappings: a gambling casino, 24-hour room service, films shown in staterooms, and an indoor-outdoor pool. It will make the run from New York 72 times a year, connecting with a second ship that will complete the leg to Miami—in all, a three-day excursion.

For travel agents who will plan tomorrow's trips it may indeed be possible to determine the proper trip for the proper personality through the computer. After all, if computer dating has reached the college campuses, why would it not be possible to match the desire and preferences of a client with a resort and a mode of conveyance? Present computers can help to select the future careers of high school students. Tucked away in the memory of the computer are descriptions of thirty thousand different jobs. Given the talents of a student, the computer can thumb through its library and produce not merely job descriptions, but the three best schools in the southeastern part of the country that teach the prerequisite skills. It will be simple to feed the computer's memory with descriptions of hotels, resorts, costs, travel times, available package rates and other data. Videotapes could, on command, produce pictures of the place chosen.

In the future there will be less of a chance of checking into a hotel, having the clerk rummage through the card rack and turn, with that sickly smile known to travelers the world over and say, "Who exactly made your reservation? We don't seem to have a record of it." Moreover, there is less chance of a reservation's being canceled because of a late arrival. Traditionally, hoteliers have protected themselves against empty rooms by holding their bookings only until early evening. If unclaimed by that time, reserved rooms are released for sale. In the future, computerized bookings will list the guest's credit card number and if the room isn't occupied the no-show guest will be charged anyway. On the other hand, the reservation will be honored no matter what time the guest arrives.

Simplified hotel bookkeeping will start the moment a guest orders two fried eggs for breakfast. The waiter will punch the order on a keyboard that will print out the request for the chef and notify the purchasing agent that two eggs have been removed from the kitchen inventory. At the same time the charge will be entered on the guest's bill, removing that responsibility from the night clerk, whose onerous task it has always been to post charges to individual accounts. No longer will a guest standing at the reception desk have to capture the attention of the night clerk, busy at his accounting chores, in order to get a room key.

The room key itself will not be standard, either. When first registering at the hotel the guest will receive a plastic card imprinted with a code. At the door to the room the guest will insert the card into a slot. The code will be electronically "read," and the room door will open. After the guest checks out the code will become inoperative. The next guest will get a new code. The security of the hotel guest will be further ensured by devices that can react to unauthorized human presence in the room when the client is out of the hotel.

"The hotel of the future," said Curt Strand, president of Hilton International, with 78 hotels outside the continental United States and another companion chain inside the United States about to start, "will have to play an even more important role as an oasis of repose, relaxation, convenience and comfort. More services and facilities will have to have extended hours of operation to accommodate the guest whose schedule doesn't allow that swim, meal or haircut between the hours of 9 A.M. and 6 P.M. The real problem in travel will be the increase of hassle, particularly in going to and from a destination."

Some airports, such as Kloten in Zurich, are already linked by rail service to the nation's entire railroad system, and thus with any point in Europe. Boston has subway service right to Logan Airport connecting with comfortable buses that make the rounds of the airline terminals. Montreal's Mirabel, maligned as a white elephant by many, is viewed by Jean Drapeau, that city's mayor, as the eventual terminus for all means of ground transfer. Drapeau, the civic visionary who was the catalyst behind Montreal's Expo '67 and the summer Olympics of 1976, sees Mirabel as the airport for much of the entire northeast corridor of the United States and Canada, with passengers being transhipped by high-speed rail, short-takeoff-and-landing aircraft, helicopters, short-distance buses and city-linked subway. What passengers want is shorter distances to planeside, which Mirabel accomplishes, and easy access to other transport. The jet aircraft may move at 600 miles per hour, and SST at double that speed, but what counts is the elapsed time from home to ultimate destination and the delays, waits and connections may well reduce the 600 mph. to an average of about 60 mph. The futurists look to the abolition of ticketing at airports. Tomorrow's ticket may well be written by the traveler himself at his own office. With distance between airline counter and departure gate cut down to a few steps, passengers will be invited to carry on their own baggage, eliminating checking and retrieving.

Those intermediate inconveniences will tend to dissolve as we move to new technologies. It is not in man's nature to stop at Mach 2, or 1400 mph., supersonic transports, or even at Mach 3 SSTs. Before the SST had even appeared in the skies over the Atlantic some missile and space specialists were talking about a one-stage rocket with four decks for passengers, each deck equipped with 43 individual couches. The rocket would float at 17,000 miles an hour, 125 miles above the earth, over the so-called thermal thicket where SSTs fly. Coming down, it would land softly, on its base, in the manner of lunar landings.

In the rocket age, Los Angeles to Honolulu becomes an 18-minute trip, and no place on earth would be farther than a 45-minute ride. However heady those concepts may seem, it must be recalled that—Jules Verne and other visionaries excepted—we have rarely been able to comprehend the miracles that will bubble to a boil in tomorrow's laboratories. In the 1790s, Americans off on a vacation were crammed in stagecoaches and bounced over rutted wagon trails as they made their torturous way to rustic medicinal springs. Two hundred years later, which is no longer than the time that separates Joan of Arc and Cardinal Richelieu, or George Washington and Jimmy Carter, they could well be off in rockets for a week on the moon or a visit on the other side of the earth, less than an hour away, all at 17,000 miles an hour.

What a way to go.

Index

312